The Florida Keys, dawn

U.S. 1 from above the Florida Keys

Street entertainer, Key West, Florida

Ernest Hemingway House, Key West, Florida

Street life, Key West, Florida

Sunset, Key West, Florida

Craig Key, Florida

Christmastime,
South Miami,
Florida

South Miami, Florida

Miami, Florida

Market on U.S. 1, Miami, Florida

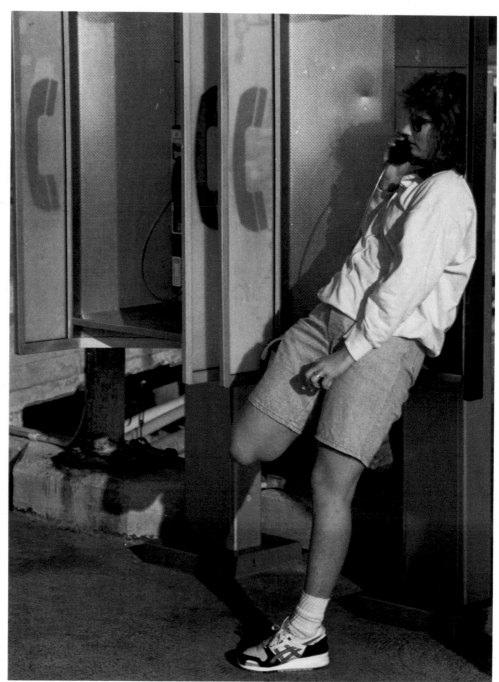

Collect call,
Fort Lauderdale,
Florida

World Gym,
Fort Lauderdale,
Florida

Sunrise, Daytona Beach, Florida

Sunset, Daytona Beach, Florida

Okefenokee Swamp, Folkston, Georgia

Okefenokee Swamp, Folkston, Georgia

A & B Auto Parts Co., Waycross, Georgia

Jim Bennett, proprietor, A & B Auto Parts Co., Waycross, Georgia

Baxley, Georgia

Town Hall, Baxley, Georgia

Onion pickers, Vidalia, Georgia

Sean Barker in his yard after Hurricane Hugo, Camden, South Carolina

Moncure, North Carolina

Railroad trestle on old U.S. 1, Moncure, North Carolina

Weyerhaeuser plant, Haywood-Moncure, North Carolina

Barn, McKenney, Virginia

Gas station, South Hill, Virginia

Parking lot at sunrise, South Hill, Virginia

Late dinner, South Hill, Virginia

Site of Stonewall
Jackson's death,
Guinea Station,
Virginia

Goalposts,
Washington, D.C.

Jefferson Memorial,
Washington, D.C.

Horse farm,
Hickory, Maryland

Baltimore, Maryland

Storefront,
Baltimore, Maryland

Baltimore, Maryland

Oxford, Pennsylvania

Moonrise, Concordville, Pennsylvania

Mummer's Day Parade, Philadelphia, Pennsylvania

Mummer's Day Parade, Philadelphia, Pennsylvania

Palace Hotel, Jersey City, New Jersey

Manhattan observed from U.S. 1 on George Washington Bridge

Fourth of July, Rye, New York

Playland, Rye, New York

United Methodist
Church, Mamaroneck,
New York

Boatyard,
West Brook, Connecticut

Mannequins,
Greenwich,
Connecticut

Stew Leonard's shopping mall, Norwalk, Connecticut

Jewish cemetery, Waterford, Connecticut

General store, North Kingston, Rhode Island

Faneuil Hall, Boston, Massachusetts

Bus stop, Boston, Massachusetts

Boston from Nahant Harbor, Nahant, Massachusetts

Abandoned development site, Lynnway, Massachusetts

Hilltop Steak House, North Saugus, Massachusetts

Garden at McDonald's, Portsmouth, New Hampshire

Hampton, New Hampshire

Hampton, New Hampshire

Van Buren, Maine

Road off U.S. 1, Kittery, Maine

Ogunquit Playhouse, Ogunquit, Maine

L.L. Bean, still open, 10:00 P.M., Freeport, Maine

Abandoned sloop,
Wiscasset, Maine

U.S. 1 and railroad
tracks, Brunswick,
Maine

Automobile graveyard, Thomaston, Maine

Sammy Saunders,
Calais, Maine

Old house, Pembroke, Maine

Madawaska Drive-In, Madawaska, Maine

End of the road, Fort Kent, Maine

U.S. 1

VILLAGE GR

ST. MARTIN'S PRESS · NEW YORK

U.S. 1

AMERICA'S ORIGINAL MAIN STREET

ANDREW H. MALCOLM

PHOTOGRAPHS BY

ROGER STRAUS III

Design by Doris Borowsky

Library of Congress Cataloging-in-Publication Data

Malcolm, Andrew H.
U.S. 1, America's original main street / Andrew H. Malcolm,
    Roger Straus III.
            p.   cm.
        ISBN 0-312-06480-2
    1. United States—Description and travel—1981— 2.
United States—Social life and customs—20th century.
3. United States Highway 1. 4. United States Highway 1—
Pictoral works. I. Straus, Roger II. Title. III. Title: U.S. 1.
E169.04.M35   1991
917.4—dc20   91-19824
                                                    CIP

First edition: October 1991
10 9 8 7 6 5 4 3 2 1

For Doris and Connie

CONTENTS

ACKNOWLEDGMENTS

I am grateful to the staff at St. Martin's Press, most particularly to Robert Weil, for conceiving this project and then nurturing it every mile of the way. My special thanks to Karen Gillis, Doris Borowsky, and Andy Carpenter for making a beautiful book.

I am told that collaboration can be tricky. This one was bliss. Sharing the road with Andy Malcolm enriched my vision of America and reminded me of the joy of friendship.

Most of all, I wish to thank the citizens of U.S. 1. In spite of all the vicissitudes of late-twentieth-century life, they display an openness and generosity that gives one hope.

Roger Straus III
April, 1991

As the last member to join a project of this scale, I have to thank everybody. So, thanks, everybody.

Now that it's done, now that the last conversation is over, the last 7-Eleven Diet Coke sipped, the last burger downed, the last word written, and the last mile driven (for this book, anyway), I'm going to miss it.

The doing of every book takes on a life and a personality of its own. All are demanding, but some are grumpy, some are moody. This one was friendly. My kind of work. Go places. See things. Talk to people. Have fun. Then get up early in the morning by yourself and write about it while the sun comes up. (I also don't mind getting paid for that, either, but don't tell anyone.)

I can do without the uncertainty of balky motors. But I have always liked the serendipity of traveling, more specifically, driving, not knowing what is around the next bend—on those roads that still have bends, anyway.

I had not realized until I got deeply into this project how much our country has changed—some for the better, some for the worse, some sideways, I suppose. So I learned a lot doing this book—about this country, its history, its people, myself, and a new close friend. At the risk of exchanging syrup with my partner in this project, I must say that for someone who lets his lenses do the talking, he for once found some perfectly appropriate words before about friendship. As an airline pilot, who spent weeks at a time traveling the globe, once said to me, "You spend precious time with your loved ones and then, after several weeks, you have to go home." In hopes that our team does similar projects in the future, I shall keep our CB names secret for now.

There are many other people I need to acknowledge. I can't mention them all because St. Martin's Press is known for tight controls on the paper budget. But I must name some. Bob Weil, of course, who keeps telling me he read my *New York Times* articles while in high school, which really makes a middle-aged man feel, well, middle-aged. My wife, Connie, who put up with my absences (or wait, maybe . . .) and provided, every morning as usual, such invaluable insights through her editing of the previous day's production. Also, she's a good buddy—after the first cup of coffee.

Although they did not accompany me on this journey from winter into summer, I must thank our children—Christopher, Spencer, Emily, and Keddy—whose fannies have patiently ridden through more automobile miles than we can count. They are good travelers, and on every trip they let me tell my father's joke about the lost Indian. I was also excused from several games of license-plate bingo.

Although this is our book, I must acknowledge the contributions drawn from many, many other sources. These would include Bruce Dale, who has made so many trips on this old road, Larry Purdom, John Carter, Matt Wald, James LeMoyne, Chris Hedges, Chris Malcolm, Francis O'Neill of the Maryland Historical Society, Joan Fradley at Brown University, Patricia K. Marshall of Mobil Oil, Ed Rider of Proctor & Gamble, the Virginia State Library and Archives, Liz Triplett of the Richmond Public Library, and Sara Reichley and the Providence Athenaeum. Countless others were most helpful in talking with me and sending published articles, old and new, pamphlets, literature, tips on other sources, and even neighbors' phone numbers; without such cooperation and thoughtful guidance, it would have been impossible to make this an entertaining journey down the road and through the years.

Richard Weingraff, historian and walking highway encyclopedia at the Federal Highway Administration, requires a special nod of appreciation.

In effect, this book draws on a lifetime of travel and experiences, and I have lost track of where I learned what. But I do want to mention several books that have proved particularly helpful now and in the past. They are: *The Americans: The Democratic Experience* by Daniel Boorstin;

The Road Is Yours by Reginald M. Cleveland and S. T. Williamson; *The Old Boston Post Road* by Stephen Jenkins; *The Old Post Road* by Stewart H. Holbrook; *Open Road: A Celebration of the American Highway* by Phil Patton; *Stagecoach East* by Oliver W. Holmes and Peter T. Rohrbach; *Coaching Roads of Old New England* by George Francis Marlowe; *John Paul Jones: A Sailor's Biography* by Samuel Eliot Morrison; *Historic Americans Roads from Frontier Trails to Superhighways* by Albert C. Rose; *Historical Collections,* vol. 37, by the Danvers Historical Society; *The Prize: The Epic Quest for Oil, Money & Power* by Daniel Yergin; and *Unobtrusive Measures* by Eugene J. Webb, Donald T. Campbell, Richard D. Schwartz, and Lee Sechrest.

Above all books, I am indebted to the anonymous authors of *U.S. One: Maine to Florida,* who, as members of the Federal Writers' Project of the Work Projects Administration, blazed the path for writers down this same highway some sixty years ago.

Finally, I want to make a sentimental acknowledgment to my parents, Beatrice and Ralph Malcolm, who took me along on so very many very long car trips so very long ago and first ignited that curiosity that I draw on every day of my life. They are off on another journey now, but somehow I feel as if they're always along in the backseat.

Andrew H. Malcolm
June 1991

INTRODUCTION

IT BEGINS IN THE WOODS and ends by the sea.

And in between are 2,467.7 miles of people, places, and stories.

U.S. 1, the highway, is a former Indian path that now runs, or rather walks, the length of the United States from the northern top of Maine to the southern tip of Florida. But besides fourteen states, this wandering route also spans American history from the Indian wars to Star Wars, from moccasin-clad explorers in deerskin to thong-wearing tourists in acrylic plaids, from smoky stagecoach taverns to microwave-beeping 7-Elevens.

A very long time ago, this route began to bind some independent-minded, isolated colonies together. Not so very long ago in time, 1 (the primary digit was not chosen by chance) was *the* major highway that began to unite Americans as a modern nation, to give an optimistic, unhurried people a sense of the national whole. They took their horses and buggies, and later their horseless buggies, and set out, as restless Americans always do, instinctively to explore, to see, to meet, to wander, and to learn, and most of all simply to go somewhere on wheels. That national penchant prompts Americans to spend $350 billion on such personal transportation every year.

U.S. 1 was both the means and the end: a road famous enough to be an attraction in its own right for travelers newly enfranchised by the motorcar and the means of reaching other places, as well. In driving the road, they saw these fresh new places full of novelty and promise and excitement, places they could see only dimly in their imagination. It was a wondrous experience in those innocent days of the tourist home and the motor court, when roads were for wandering, when people were neigh-

bors and paid cash, and when Americans still suspected the best of each other until proven wrong. It was a simpler time, this first half of the twentieth century, when fathers did the driving while thoroughly tanning their left arm, when mothers sat to their right, picnic basket at hand. The children perched in the back (they, too, had not yet heard the expression *single-parent family*), their ears and minds not yet conquered by individual headphones made in Taiwan whose soft earpieces and hard music made isolation so easy and comfortable and seemingly safe.

Today, along all those hundreds of miles of U.S. 1, there remain but five roadside picnic sites.

In the early days of motoring, it was considered a pastime in itself, not a means to the end of an errand. Driving often involved motors of uncertain reliability and roads of reliable uncertainty. That was a time when windshield wipers slowed with the cars at each red light, when radios were not called sound systems, when one speaker seemed sufficient. Even in more recent years—my childhood, for instance—it remained a time of innocence that could produce wondrous moments when the big truck blasted its air horn in response to my waves, when spotting a license plate from our home state was worth a friendly nod, when car windows were open and warm breezes blew, when the expression *fellow travelers* was not bad.

Every hill, every stop, every turn on those two-lane roads held the promise of newness, of excitement, of the unknown, because no one had yet seen everything there was to see during the brief breaks between the commercials on a screen in their very own air-conditioned living room with the doors safely locked. There was romance and mystery and suspense and surprise on that highway that could produce a vivid natural high of living that only words, not videotape, can capture.

Americans by the wheeled millions came to this premier road, this first national highway that ran down the prominent, populous spine of an unpredictable land on the verge of a half century of historic change that would both inspire and convulse the world's oldest surviving democracy. As they moved up and down these two simple lanes of pavement over all these years, the people changed—and so did the road. But not as much as the folks who didn't drive it so much anymore. What that road and that roadside have become reflect much of what those people have become, what they value, and what they no longer prize.

What was left behind from all these years, besides fallen Burma-Shave signs, crumbling Styrofoam wrappings, and memories as faded as the first thrills of childhood, is an old king of a road.

Today, few people know this road beyond the local dry cleaner's. Today, most long-distance travelers who choose to stay ground-bound opt for the Interstates, those efficient, high-speed cocoons of concrete copied from Hitler's autobahns that consume forty-five acres of land in every mile. Such highways allow Americans to rush from one place to a second place that looks just like the first with maximum speed and minimum time (E-Z Off-On)—and with minimum involvement with each other and the roadside world beyond the federal fences.

As our passion for speed increased, it literally changed the shape of our lives—and our thinking. Everything has become streamlined—from our buildings and the grilles on our cars to our cooking and our relationships. Time like money became something to spend instead of invest. No time to prepare a full meal? Invent the microwave. Pop in a precooked platter. Nuke it. Chow down. No time to loiter in a graceful lobby lined with woods and warmth; coat it with chrome and glass, take out the overstuffed chairs and the friendly old elevator operator, keep those people moving with the heat-sensitive elevator buttons and the TV cameras silently watching from the ceiling for safety's sake.

No time to dine? Create a fast-food industry with revenues of over $111 billion per year with the spending on burgers alone ($30 billion) exceeding the GNP of many countries. Turn any meal into another refueling stop with a drive-thru window and on-site mini-radio net so customers can stay in their familiar cars, even though the express lane often is less express than the line inside. And like as not, you get the wrong sandwich, anyway. Now there are drive-thru lanes at liquor stores and even funeral homes; we seem to save time, and that's the important thing in these modern times, when *TV Guide* is considered serious reading and there's no time to dial the phone. So invent the speed-dialer ("Instantly dials 500 numbers at the touch of a button"). And overnight package delivery is too slow, so we must invent the fax.

Streamline the front of cars, even the windshield wipers, to maximize airflow and minimize fuel flow. Ditch the hood ornaments and the tail fins and the social modesty that cluttered yesterday. Streamline the service out of service stations. Shed the inhibitions that adorned life but surely slowed it down. Get the colors and the shapes that are hot, Hot, HOT!

There's no more forever today, just the now. Tune in the radar detector. Set the cruise control (C.Cntl.). Punch up the CD for the aurally fixated. Or the cellular phone. (And if you just want to appear hectic, there are very genuine-looking fake car phones.) Set the Air Cond for Max Cool. Push one button to adjust the mirrors (CAUTION: OBJECTS IN MIRROR ARE CLOSER THAN THEY APPEAR). Punch another lever to close the windows, a third to seal the doors. Flick the gearshift to D and you're off, all the way around the block to tap your P.I.N. into the bank machine at the convenience store.

During one recent journey on a warm Sunday afternoon in South Carolina, I backed up to the gas pump because another customer was already there. I leaped out and frantically began to fuel my tank, just as he did. Around the five-dollar level, we both looked up at each other in shared amazement and humor. "Why are we in such a hurry?" I asked the stranger.

"Damned if I know," he said. We smiled. Then we

hustled to the bulletproof cashier's cage. He got there first.

I probably have traveled a couple of million miles around our land over the years, much of it by land by choice. When I was a child, back in the forties before man broke the sound barrier, by car was the way most people traveled. The only reason to fly was an emergency, like a family funeral far away. My father drove forty-five minutes to work each way each day before that was so common. When he retired, he said, he wasn't even going to have a house. Everything he owned would be in the car and he would wander anywhere he wanted anytime.

He never did that, but I think the anticipation of such freedom helped get him through some long ordeals in the office. When I sat in algebra class and that kindly, caring man named Mr. Smith droned on about factors and equations, my eyes would often drift to the left out the window to the parking lot, where my mind would turn the ignition key, and off I'd go to so many wonderful places so very different and so very far away. It never seemed strange to me that in my youth many American cars such as DeSoto, Hudson, and Cadillac were named for explorers. Even the bulbous Rambler connoted wandering.

In my childhood, I was never happier than in the car with Dad going somewhere new or even somewhere familiar on an old road. There are some roads that I could drive blindfolded; if I never again navigate I-80 around the bottom of Chicago through Gary's guts, that would be just fine with me; it was more interesting the old way right past the steel mills with the bright glow seeping through their gates.

Nowadays, I am never happier than in those early hours of a fresh day when I head out somewhere new, not knowing where I shall sleep on that coming night. It's exciting. It's serendipity, just wandering about to see what there is to see, to hear what there is to hear, to smell and taste what there is to smell and taste. To collect all kinds of regional arcana and trivia. To meet people I've never known. And to safely ignore whatever I want. My destination is wherever I am.

I wandered in the woods when I was little in the days

of B.L. (Before License). Even in my room, I wandered the world on the dial of my shortwave radio. For a lot of years, I have been paid to wander the byways of America and then write about it for readers who may not remember, or ever have known those ways and byways.

Thanks to technology, including this word processor these words are flowing into, we have gained awesome convenience. But there is a cost: We are losing, among many other precious things, the serendipity of life. Computers require clear instructions and precise search terms in a language that fits their logic. Our busy lives, too, have so many constraints; we have time only for sure things. No more time for chance encounters. I remember in my childhood once, just once, asking my grandmother how to spell a word. That kindly woman, a prairie pragmatist who taught a generation of runny-nosed farm kids in a one-room schoolhouse many miles from town, looked down at me with her one good eye. "If you don't know how to spell a word," she said, "then look it up in the dictionary." Which made marvelous sense until I discovered that in order to look it up in the dictionary, I needed to know how to spell it. Now, when some computerized information-retrieval system informs me in its maddeningly superior tone, "Improper Search Term," I feel like typing back, "If I knew what I was looking for, I wouldn't need you." Disconnect.

Americans have always been searching for something. Even a century and a half ago, Alexis de Tocqueville noted how restless Americans were. From the Minutemen to the mountain men, from the Sooners to the Okies, from beatniks to Rust Belt refugees, so many have wandered the byways. "See the USA in your Chevrolet," Dinah Shore sang in the optimistic fifties. And so millions did. Now, the same car company advertises "The Heartbeat of America." It isn't clear what that is, but the nanosecond snippets of Americana sure feel good for a generation raised on MTV in an age where a minute is an eternity, even for parents just passing through the room.

For a while in Colonial times, the only byway was the seaway and the rivers as far upstream as tide and tiny boats could go. Which is why so many early communities sprouted at the first falls or rapids. Then, slowly the newcomers began using the foot-wide footpaths of the Indians. They have been widening the roads ever since. The largest public construction project had nothing to do with the pyramids or the wall in China. It was the 43,500-mile-long system of Interstates launched in the mid-fifties by President Eisenhower, a Kansas native who had personal experience with rugged roads. And if an initial rationale for these multilane, multibillion-dollar behemoths was civil defense—after all, in a nuclear attack everyone would have to get out of the targeted cities as fast as possible, wouldn't they?—then the actual result was to put more distant destinations within vacation reach of Americans.

In fact, with the brief exception of the World War II years, Americans have given birth annually to many more cars than babies. They spend more on this vital mechanical product than they do personally on food, shelter, or education. Every year more than 164 million licensed drivers steer their 180 million vehicles down nearly 2 trillion miles of highway. Most of these travelers get there safely—although more Americans have been killed on their highways by Americans than have perished at other enemy hands in all the nation's wars. Still, never before has driving been so safe.

Like the rivers before them, many local roads have become transportation backwaters, often crowded and polluted, overlooked and taken for granted with aging, discarded buildings and fading memories nearly lost to time in the unproclaimed race to modernize everything.

Although the freeways, those high-speed urban maelstroms, seem the most frightening, it is the local slow lanes that claim the most victims, because they remain the heaviest-traveled, as they have been throughout history. The Romans, it has been noted, used roads to spread their culture. For Americans, it often seems, roads *are* their culture: tacky and beautiful, narrow and wide, straight and curving, crowded and vacant, smooth and potholed, scenic and cluttered.

Roads have been crucial throughout history, and not just for conquering armies. Until the detailed development of roads such as U.S. 1 and its predecessor paths, life was pretty much confined to your own town or region. Indian runners could cover more than one hundred miles a day. First, roads and horses and then cars came to liberate people physically and mentally from places and lives, to open the possibilities and to create a great washing to and fro on the cement ligaments that tied town to town, city to suburb, and now suburb to suburb and mall to mall. Where roads crossed roads or railroads, historically, communities grew. There are many times when these slow lanes seem to link gas station to gas station. But in between the Qwik-Stops, Fast-Marts, and Jiffy Lubes, there remains a rich fabric of people who remember, of stories to tell, of thoughts to think, and of lessons to learn.

U.S. 1 is a modern name for an old road, the Boston Post Road and the Kings Highway, although no king ever rode down it. U.S. 1 in all its previous forms and names and varying routes was America's first Main Street, connecting the yet-to-fade metropolises of the Eastern Seaboard even before 1925 when the federal government began collecting pieces of local roads and numbering them regally. There was a brief controversy at the time; many people objected to a central government erasing evocative local names, which had been good enough for 150 years, and substituting impersonal numbers. Just like a bureaucracy. It would take a few more generations to expunge likewise the evocative names of telephone exchanges like BUtterfield 8 or OLympic 3. But efficiency won out. And now, thank goodness, every American surely knows that when a highway number is even, it goes east and west, and when it is odd, it goes north and south.

Highway 1, which actually goes more northeast-southwest, remains an identifiable road through much of a crowded corridor that contains both unspoiled countryside and about one-third of the United States' metropolitan population.

Nineteen of the country's 180 largest cities are on U.S. 1 from No. 1 (New York City) to No. 177 (Stamford). And so is the nation's worst city for murder and infant mortality (Washington, D.C.).

Although no one in their right mind—no one, that

is, who isn't writing a book on U.S. 1—would ever drive the length of old 1 straight through today, it is possible. The precise route has shifted countless times. As a living entity, the road continues to change daily, as do the people who drive it, walk it, farm it, clean it, watch it, and ignore it.

The road is now unofficially divided into countless local segments jammed with hurried commuters, delivery trucks, fast-food joints, carpet stores, fast-food joints, muffler shops, fast-food joints, and both mini-malls and major malls. But U.S. 1 also has pristine parts where trees crowd the shoulder, horses watch the traffic, and families collect the litter that grows faster than dandelions. And all along, the historic road also has the colorful memories of many people and places left behind to speak softly in this noisy era of a quieter time and a gentler way.

Today, from the shoulders of U.S. 1 near its top, where no one with a working automobile would ever stop on those blustery days of winter when the highway is open to humans, you can see Canada on the left just across the river in a different time zone and another world. At the other end in the tropics, some sunny days a million people stand on its shoulders, shading their eyes with hands to watch an immense rocket rumble away toward another planet.

In between, U.S. 1 sneaks through woods, where old trees groan in winter winds. It zips by beaches, where people sigh in delight at the thought of not being somewhere else. It twists through some of the prettiest, most under-stated countryside with possibly the tightest ratio of people per golf course in the world. In waddles through the troubled, flatulent streets of our largest, aging cities—Boston, New York, Philadelphia, Baltimore, Washington, Richmond, Jacksonville, Miami. Once, the ragtag Continental army straggled along what would become U.S. 1 (WASHINGTON MARCHED HERE. ALSO LAFAYETTE). Ben Franklin set mileposts along its untamed shoulders for the mail riders and stagecoaches that rattled by. Early Supreme Court justices rode the circuit on the road. President Washington toured the nation on it and had some time-consuming trouble (WASHINGTON'S HORSE WENT LAME HERE).

Along one part of 1, they massacred fleeing Indians in the name of civilization. Along another, possibly 100,000 men killed each other in an uncivil civil war, though try as they might, they never did find the parts of all of them. U.S. 1 links all three cities—New York, Philadelphia, and Washington D.C.—that have been this country's capital, and a little Pennsylvania town that nearly was. It even passes by a haunted inn where new suburbanites now dine, unaware. In most places, U.S. 1 is well marked; in others, it's lost. In some, people like Glenn Maura, who patrols it, aren't even aware of its name. People jog on 1. They sell cars on 1 (and used trucks for the latest insurrection in Central America). They walk dogs on 1. They tell (or lose) fortunes on 1. They sleep on 1. They hunt tourists on 1, oh, how they hunt tourists on 1. And they tattoo thighs on 1.

This book is not a traditional street guide or an encyclopedia; the photographs and the text provide a combined portrayal of an aging region undergoing significant change. Our journey—sometimes as a team, sometimes apart—was a very eclectic one up and down the old road. We talked to old-timers and newcomers, to those who've spent a lifetime on the road's shoulders and those who do it every rush hour. We talked with teachers and wardens, with farmers and bureaucrats, with candy makers and gardeners, with cabbies and caretakers, vets and cops, nurses and barbers, fishermen and beekeepers.

It was a happy time and a sad time, a beautiful road and a filthy road, an inspiring road and a numbing one. As always, the price of going somewhere is leaving somewhere else. Our society is leaving the slow lane of old U.S. 1 at a pretty good clip. And we hope that as you make this journey with us, you, too, will come to see not only where we've been but will gain a peek perhaps into where we might be headed down the road.

PART I

MAINE TO CONNECTICUT

The Top

FROM THE AIR THERE, it looks like a path, which it once was.

U.S. 1 rises from the earth as West Main Street in Fort Kent, Maine, to begin its sinuous trek to the distant south. A sign in front of the duty-free shop announces on one side the start of the historic road and the end on the other. Not for the first time in history, a Chamber of Commerce committee thought the sign might become a little tourist attraction, a kind of photographic destination where travelers look at the camera and point at the words, like successful hunters with their foot atop their fallen prey. It works—at both ends.

Although the highway—others might accidentally call it a road—twists even farther north before succumb-

ing to the sun and the South, Fort Kent's vital land link to the world cannot appear more arctic on blustery winter days. The winds blast down from Canada, one-hundred yards away. New folks choose to brace themselves at the self-serve gas station, where one row of pumps dispenses fuel in gallons and U.S. dollars and another is set for liters and Canadian dollars. The winds grab the smoke wisps from chimneys and exhaust pipes and whip them away into nothing within a second. Waves of snow—part of the ten-foot annual harvest of the stuff—scud across the glazed pavement, where the ruts of ice have been crunched by passing four-wheel drives. The bank sign says it is minus two. But any exposed skin puts the temperature at something closer to minus twenty.

Fort Kent (Edward Kent was a nineteenth-century governor and the fort was to guard against an invasion of Canadians that never came in a nineteenth-century border dispute) is a practical, proud, working-class town. It is the kind of smallish community that has the thought and takes the time to erect a road sign: CAUTION DEAF PERSON. The residents make time for overtime but have little time for fashion beyond function. Here, Stan Albert over at the Village Squire has difficulty convincing men to wear anything more outlandish than a striped rugby shirt. For a while, Stan thought his clothing store was leading a fashion revolution in the northern woods; the women, who always are more adventuresome, began buying colored underwear for their men, and shiny trousers and blue jeans with buttons on the fly.

Stan began envisioning all kinds of fashion possibilities walking off his racks from both his American and Canadian customers, who wear their new Nikes home to avoid any duty at the border. He imagined that the cash register would be ringing constantly behind the frost on the display windows at the Village Squire. And Stan would soon be basking on some southern beach every February.

However, as sure as the snows can start in October, a few days later the wives, shaking their heads and clutching the store's crumpled paper bags, began reappearing at his store on Main Street. They were returning the high-fashion items Stan had ordered with such high hopes—not so much the colored underwear, which the husbands could wear for the wives without any of the guys knowing, but the shiny pants and the buttoned-down jeans and shirts.

They'd stick with their flannel plaids, *merci beaucoup*. And their work boots. When the Fort Kents of rural America were isolated, back before the roads were paved and expanded, they had their own hard-bitten fashion rules. And there was no use for tasseled loafers. The men, mostly French-speaking Acadians with names like Thibodeaux, Levesque, and Nadeau, were descendants of the colonial French expelled from Nova Scotia for not swearing allegiance to the British throne.

Here in Fort Kent, they swore plenty, in two languages. They went into the woods in the fall, right after the potato harvest when the ground was freezing up the muddy paths. It was a grim, raucous, sweaty, crude, profane, physical life among the trees, which were upward of six feet in diameter and took a day apiece for two men to fall with their crossbuck saws. The workers in their cluttered camps fueled themselves for the long days' labors with mountains of biscuits and platters of beans, fatback, and salted cod.

Few people traveled much to other parts of Maine because it was hard and time-consuming, there still were highwaymen lurking about, the state was too damned big anyway, and there didn't seem much need to travel, unless there was a war somewhere else. As late as the 1920s, the Dubois brothers lived but thirty miles apart after their weddings and never saw each other again. Of course, everyone went back and forth to Canada; that other country was local. Fort Kent folks still go over to Clair, New Brunswick, for ice-skating and dancing; the Canadians come over for cheaper groceries, kids' clothes, and Nikes.

Families were self-sufficient in the old days, growing their own foods and making their own garments. And the families were large, very large, in part because there was no television to postpone bedtime and in part to handle all the work and the misfortunes of life. Stan had three brothers and a sister; his mother, Therese, was the youngest of eighteen. Roger Paradis, the future local history teacher, was one of seven children born to Mary and Fred, though only five survived infancy.

When springtime turned the frozen roads to mush, the logs could no longer be dragged to the sawmills by the ox teams, and the sap was rising everywhere. The men trudged from the woods to their home to plant seeds in the fields and elsewhere, which is why just about everyone in those parts in those days was born shortly after New Year's.

There were roads, to be sure, though none so formal they were named or numbered. The path along the river was simply *chemin du bord.* There was the St. John River, too, the heart of the peaceful, modest valley that Henry Wadsworth Longfellow wrote about in *Evangeline,* in 1847, so that so many generations of schoolchildren could sleep through English class. And the Canadians, who still aren't so much a different nationality as they are a different branch of the same families who populate the American side, had a railroad early on. When the American boys went off to the Civil War, most of them not to return, they journeyed for days by foot and wagon down through Frenchville and Madawaska to the mustering point in Caribou, about forty miles distant. It took nearly three weeks for word of Lincoln's assassination to finally reach the top of the country. But by the time Fred Paradis left the woods to go bayonet other fathers at Château-Thierry, Argonne, and Belleau Wood, over fifty years later, the Americans finally had a railroad. That's when Fred learned to read, too.

One of the first automobiles was built and run in 1892 by the Duryea brothers not too far away, by Interstate standards, in Springfield, Massachusetts. But it took a while for that machine to develop a presence around Fort Kent and for that presence to elevate the importance of roads. Among the first to use them both commercially were the rumrunners from Canada, traders who smuggled in the illicit liquids from France and Quebec in every conceivable cavity in their machines and clothing. Many a table in local homes had a hollow leg where the booze could be poured hastily upon the arrival of a revenuer. There was the local woman who offered her favors to the local policeman on every evening when her rumrunning husband was out; no one is sure what the primary motivation was, but it sure helped to keep track of the law.

Fred Levesque from neighboring New Brunswick was reputed to be the best exporter. He had a fleet of some seventy Model T's and other vehicles shuttling back and forth between Canada and the Boston States, as some Canadians still call New England. His drivers seemed to have a lot of relatives they needed to visit to the south down the dirt highway. One day, after some new American customs officials at the bridge had grown suspicious, a convoy of prominent Canadian church people appeared at the Fort Kent crossing in dozens of vehicles. They were enroute to the funeral of an important bishop in Portland, they said. They were dressed like priests and nuns. And so the agents let them pass, figuring the cars appeared so heavily laden because of all the people.

By World War II, the demands of distant markets for people, potatoes, and wood made the main road out of Fort Kent busier than ever. So it was paved well beyond the town's forested limits, where it was called the Frenchville Road because that's where it went. But a road that goes out also goes in. And up the main valley road from the wider world came new people, new machines, and new ways that seemed to make life easier but also made self-sufficiency more difficult, forever changing local life. There were telephones that cut down on local socializing, and televisions and one-eyed satellite dishes that plucked invisible signals from the frigid air and seemed to keep folks home more until the later advent of VCRs sent them out, briefly, to rent a movie. Roger Paradis believes such things have weakened a generation's imagination; he's certainly seen weaker student memories and a decline in student writing.

Things like dining out are more expensive for families now in an era when virtually everything from vegetables from Florida to Stan's new flannel shirts from Pakistan are trucked in from outside over the molar-jarring frost heaves along old 1. They're still mighty friendly at the local restaurants, but the other day at Debois's place, the waitress, another working mom, showed no embarrassment at running out of mashed potatoes by lunchtime.

Roger Paradis originally thought all these changes might be good. Conveniences certainly made life more, well, convenient. And how could anyone argue, for instance, against the system of dikes that went in during 1977 and almost overnight halted the annual scourge of spring flooding? Perhaps an easier local life would help keep the young people and families from fleeing the rural areas in search of ease and work elsewhere, leaving the teen population so depleted that, come the early, frigid falls now, the Fort Kent Warriors compete with the arch-rival Madawaska Owls only on the soccer field.

It hasn't worked out as Roger hoped. While a few tourists come to hike and canoe—and more are welcome (hint, hint)—most of them head for the seashore farther down 1, which isn't all that less frigid even in summer when people spend a lot of time sitting on the beach preparing to enter the water and then sometimes go wading up to their ankles. Like many old rural communities along Route 1 and elsewhere across the country, too many locals have left. In 1940, the Fort Kent area had 5,363 residents. By 1970, the number had declined to 4,500. Today, it's still slowly decreasing and the town itself has slipped back toward 4,200.

Roger's suspicions were first ignited when some old-timers said they could not recall enduring any floods in their youths even without the dikes. He got to thinking about this during his regular autumn hunting hikes, though his thoughts were frequently interrupted by all the other passing hunters. They'd tromp by in their L. L.

Bean camo outfits enroute to or from their suspiciously clean four-wheel drive vehicles parked on one of the many logging roads that now honeycomb the entire area.

The woods had long seemed closer to town. Now they seemed to be getting smaller, the clearings larger, and some erosion more prominent. Of course, with the advent of the chain saw, removing trees was much easier; a man could take out over a hundred big ones a day now. So they did. And with the short growing season this far north, it takes so much longer for replacements to grow back. As one result, there weren't nearly as many living things standing around to hold the hillsides and soak up the snows and rains and let the moisture evaporate slowly through their leaves and needles.

Roger wrote some major American conservation organizations about the problems he saw developing. They sent him back several pounds of materials on the disturbing environmental and social problems of clear-cutting in the distant Amazon jungle, where none of the funds' donors were involved.

Roger dug up some figures on hunting. In the early 1950s, the annual deer-hunting season produced takes of 48,000 to 50,000 animals. By the 1990s, three times as many hunters with their four-wheel drives and modern weapons were killing 19,000 deer. "You destroy the habitat," he said, "and you destroy the inhabitants."

An Otherwise-Perfect Name

THERE IS NO CARIBOU HUNTING AROUND Caribou. This is not to protect those regal creatures from beer-drinking bozos with large guns and small heads. This is because there are no caribou around Caribou.

The last wild one was spotted on a sunny afternoon on Mount Katahdin in 1908, right around the time cars began taking hold of life in Maine. Since then, they've

become extinct, the caribou, that is, due to a surplus of good shooters. Several times various groups tried to re-seed a wild herd with imports from Canada, which still has so many road-free wild areas. In 1963, they released twenty-four caribou in Baxter State Park southwest of Caribou; human and wild predators soon took care of them. In 1986, another expedition shot two dozen of the creatures with tranquilizer darts in Newfoundland. They plopped them into helicopter slings for airlifts to trucks to the University of Maine, where the ones that hadn't died of shock were shot up with medicines and vitamins. The idea was to create a healthy nursery herd adorned with radio collars and turned loose in Baxter, which was becoming the Bermuda Triangle of cariboudom. All but one disappeared, which inhibited procreation. There being no more wolves around to blame, coyotes and bears were fingered, especially the bears, who are so uncivilized they don't buy licenses to kill.

Biologists said it might take a hundred caribou to reestablish a wild herd, and the private project could not afford that. Neither could the caribou, who weren't exactly volunteering to leave Newfoundland to become targets Down East. The Maine Caribou Project's final report suggested that future money would be far better spent conserving endangered species *before* they disappeared.

Hunting by humans is a big pastime and a big business in many areas. Besides guides, camp owners, and outfitters, Maine's 215,000-plus hunters pay the state $6 million in fees each year. Maine still thinks of itself as very rural and very large; in fact, the state is almost as large as the five other New England states put together. With its 6,453 square miles, Aroostook County, at the top of Maine where U.S. 1 runs some 620-odd miles above New York, is itself larger than some states. This causes many local residents to refer to remote areas like Boston as "the outside world." These isolated rural regions are also largely empty, although better roads and cars have enabled a fair number of city folks to flee the fumes and crowds every weekday and move into the country, thus becoming a suspicious species known as "newcomer."

This population seepage has created a clash of cultures along the roadside. So deeply rooted is hunting in such areas that legal action is sometimes necessary to settle misunderstandings. In the next county south of Caribou, a supermarket produce manager went out deer hunting a couple of years ago. He saw one of those nimble animals through the brush. He blasted away. He saw the deer's tail again. He shot again.

It was a real good shot. He hit the target right in the chest. But she was a thirty-seven-year-old mother of two. She had had the audacity to walk into her backyard in the fall wearing white mittens. Some people suggested it was the fault of the woman, who had recently moved to the wilds of Maine from that famous urban center of Iowa. One witness testified the hunter had rushed to his victim's side, saying, "I have shot a human being! Oh, God. Why does God allow this to happen?" In the face of this overwhelming evidence, the local jury decided the whole thing was an understandable accident and acquitted the produce manager of manslaughter.

The widower moved back to Iowa with his daughters. The hunter's lawyer said his client was so shaken by the episode that he might never hunt again.

A less violent fixture of autumn along U.S. 1 in northern Maine is the potato harvest, which follows the summer strawberry and raspberry harvests and precedes the emergence of snowmobiles. For nearly a half century after World War II, school districts in communities such as Presque Isle routinely dismissed classes for three weeks so the youngsters could help bring in the potatoes before the fall freeze. It gave the young people a minimum wage, taught them the value of a twelve-hour day, and simultaneously produced the state's largest cash crop.

Until the previous popular war, the dank summers of Maine made it the nation's proud leading producer of those dirty little vegetables. But then all those dams and canals they built out West, where the votes were growing, too, put more arid land under cultivation. And Maine slipped to fifth place! And while the crop still brought in $110 million a year, Maine's potato acreage had fallen from 200,000 to 80,000 on just 650 farms. In Presque Isle, anyway, the number of high school students involved plummeted to barely a third of the seven hundred eligible youths. Since their working parents did not get the same three-week recess, the others watched MTV during the vacation or hung around their cars out at the Skyway

Plaza, which is the first mall south of Canada on U.S. 1. A McDonald's, Burger King, Dunkin' Donuts, and Kentucky Fried Chicken are never far away on 1, in case starvation strikes.

When a new federal law prohibited those under sixteen from working on mechanical pickers, the school districts ended the potato recess for grade schools and junior highs. Then the districts took the emotional step of ending the vacation for teenagers, too. The farmers were furious, saying their taxes heavily supported the schools that should make the inexpensive labor available.

But officials were adamant. "The whole point," said Linda Pelletier, a school-board member, "is that times have changed."

They certainly have. In 1808, even before malls, ten thousand acres of northern Maine were deeded from Massachusetts, which thought it owned Maine at the time, to Captain William Eaton for his crew's heroics in vanquishing the Barbary pirates over there in the Mediterranean. Sixteen years later, white settlers became the area's newer newcomers, which caused considerable consternation among the resident hunters, who were Indians. In 1820, as part of the Missouri Compromise, Maine became a state. Without trouble, it could take even uniformed travelers three weeks to make it the 175 miles up from Bangor. There were no roads (road building was the responsibility of newcomers as part of their land's purchase price). So most settlers came in winter, not the most pleasant time for open travel, but the frozen ground created a sort of roadway.

Today, thanks to technology and modern machinery, winter is a terrible time to travel there, even just to the Social Security office. Good-sized patches of ice, which could be future glaciers if it weren't for spring, are distributed across both lanes in interesting invisible patterns that provide a stimulating challenge to almost any driver. Large logging trucks, their fallen cargoes crusted with ice and snow, blast by, blowing the white stuff, chunks of ice, and other natural litter about in their own winds. In many places, the pavement remains unmarred by such things as visible lane markings.

Veteran Mainiacs know to hang their mailboxes on chains from a post standing back from the roadside. Anything too near the shoulder is demolished when the county plow plods by, hurling some of the heavy snow off the pavement, flattening everything in its path, and burying the wounded mailboxes until the late spring under layers of ice and snow that linger in temperatures that rarely rise above freezing. Some of the houses along the roadside of U.S. 1 are abandoned, long-term or just for the winter, leaving their windows dark, their rusting TV antennae askew, and the outbuildings gaping open to the icy winds that whisk the snows inside the doorways to land and lie in tidy drifts like white shadows. Other houses are inhabited. Those are the ones with the path well-worn to the woodpile under the bright blue tarpaulin and back to the door of the home, where many dwellers tack heavy sheets of plastic over all windows for additional protection.

The plows' snowy wakes also bury most driveways, which necessitates one of two responses by residents. Either they shovel open the entrance to their drive, tossing the moist debris onto ever-higher mounds of snow and gravel that soon tower over the shoveler, obscuring the drive from the road. Or the residents shift into four-wheel drive up by the house, get going at a pretty good clip, and just burst out onto the highway right through the man-made drifts, which is fun and provides a free morning boost of adrenaline for oncoming motorists who otherwise might have to get to work without a heart attack.

A century and a half ago, settlers around Lyndon (no one knows who he was, but Lyndon was Caribou's name until just before all those animals disappeared) took turns making the fifty-five-mile runs to Houlton for the mail down what is now U.S. 1. For many years right after statehood, Houlton was the home of a federal army barracks, a kind of human DEW line before radar in case of invasion from Canada, which never sought its independence from Britain but got it anyway in 1867. On the Caribou settlers' return, every two or three weeks, the letters were passed out from Ivory Hardison's house.

Today, the road to Houlton and the eighty miles beyond to Calais remains uncongested most times, unlike most of 1, with long stretches of virtually nothing showing in your car's high beams. At all hours, the road continues to transport a mix of people and things, not all of which are legal. Not too long ago, Canadian authorities arrested two Colombians and seized about 1,100 pounds of cocaine from an airplane that landed at a makeshift airfield not far from the American border. A few months later, on the day before the pair's court appearance, Canadian authorities in Edmundston, New Brunswick, became suspicious about two cars and a van that had just moved down U.S. 1 past the Irving gas station and Rogers Electric Motor Shop and slowly crossed the narrow bridge from Madawaska, Maine (CYCLISTS USE CARE CROSSING BRIDGE).

It *was* hunting season, and hunters *do* come from all over. But even if they are at opposite ends of the same U.S. 1, Madawaska is not Miami. And Venezuelan passports had not been proferred frequently at the border crossings of Aroostook County, where, not long into their new assignment, customs officials come to know pretty much everyone local, very few of whom have names like Felipe and Juan. In the strangers' vehicles, the Royal Canadian Mounted Police found no drugs, just an Uzi submachine gun, several Russian and Israeli assault rifles, six automatic pistols, a .22 pistol, more than three thousand rounds of ammunition, a hand grenade, burglary equipment, and an electric zap gun used to stun people. These sportsmen no doubt had nothing to do with drugs.

The larger northern Maine communities along 1 such as Presque Isle have an array of local, once family-owned businesses struggling in the old downtown, places like Anne's Fashions, Brown's Jewelers, Harris Optical, a pharmacy. Inevitably, there is at least one tidy bank proudly presenting the ubiquitous revolving time and temperature sign with at least two bulbs burned out, because by the time it's dark enough to notice the dead lights, the bank officers are out in their large homes, reading ledgers or *The Wall Street Journal.* There may be a bowling alley (SPRING LEAGU S FORM NG NOW). And if there is a movie, it most likely has been divided into two or three small cinemas, because in today's world of Hollywood economics, two fifty-seat theaters with forty customers each make more money than one two-hundred-seat movie theater with eighty tickets sold.

By five-thirty on a winter's afternoon, which is an hour after dark and an hour before the popcorn machine goes to work, all those establishments are closed, while all the big chain stores in the commercial clumps on the edge of town beyond the parking meters are still open and brightly lit. The computers there with the flickering white light beams that instantly read all the outgoing labels are already reporting back to Boston and Detroit and Chicago on how many of each item passed over the counters that afternoon.

The highway beyond the towns' limits is virtually deserted by six; even the Frito-Lay delivery trucks have gone home for dinner. The brief pale blue sunset is long gone near the southern horizon. Dotting the roadside are some small communities such as Littleton, North Amity, Topsfield, and Eaton (the pirate guy, remember?). Each has a tiny cemetery where the gate is locked, although there is no fence. Some grave markers are still standing. You can tell the newer ones; they're the tombstones laid flat in the ground to facilitate mowing.

In many towns, the sweet smoky scent of wood-burning stoves drifts on the brisk winds and there, waiting for empty gas tanks, snack-food addicts, and the working mom whose fourth grader drank the last milk after school, surely stands the convenience store, which used to be called a general store until the importance of seeming to save time arrived. The areas in between the towns are sparsely settled. The vast fields and woods are silent, snow-covered, and resting. The chilled landscape is a black and white monochrome beneath thousands of twinkling stars, though at temperatures below zero, few pause to ponder. Some of the scant houses have those lonely all-night backyard lights that can tell when sunset and sunrise come without a human hand throwing a switch. Small inverted V's of wood placed over favorite bushes stand guard against the enduring weight of ice and snow. In the dark, the homes' windows glow yellow and most movement seems centered on the kitchen. The color TV screen blabs idly to no eyes in the darkened living rooms.

There are very few road signs on this part of 1, announcing, say, how far it is to anywhere else. It is a psychologically self-contained area where people figure if you're out and about, you must be local and, therefore, you know how far some place is—or at least how long it takes to get there in each of Maine's two seasons: winter and a time when the sledding is poor. And if you're not local, why would you care? Or, why would they care what you care?

For motoring couples, this kind of situation highlights one of the major differences that God bred into the genders. I call this the "Let's Ask Someone—No, I'm Not Lost" syndrome. The roles in this marital play in three acts are as psyche-set as they are in procreation. The male will drive around for an hour or more looking in vain for their destination. The female will suggest, calmly at first, that they stop and ask directions at one of the several convenience stores that they conveniently pass. "Let's ask someone," she says, all chipper.

The male dismisses the idea as certainly unnecessary and possibly preposterous. "No, I'm not lost," he says, taking the suggestion personally, like someone passing him just before the road narrows. He might also add that stopping for directions would cost them valuable time.

We've already lost valuable time simply wandering, she thinks. But she keeps that to herself—at first.

Ten minutes later: "Why don't we stop here and ask?"

He tries to ignore it, too busy changing the radio station to get the score of the game he thought he'd be watching in someone else's living room by now.

"We could stop here and ask?"

"Not necessary." He surreptitiously checks the gas gauge: plenty.

Eventually, of course, the man and woman arrive, with their gift bottle of wine and a load of rancor the size of which depends on how long each partner held out— all because of those stupid road signs that weren't.

Letha Butler Goes Home Again

THE LIGHT ON THE neat lawn sign goes on right about dusk, which comes pretty early after the leaves are gone. BELLMARD MOTEL AND TOURIST HOME HOUSE OF DISTINCTION OPEN YEAR ROUND FAMILY RATES MEALS BY RESERVATION WILLIS L. AND LETHA BUTLER, PROPS.

The "motel" part of the wooden sign is for the younger generation that cringes more than a little at the thought of staying in someone else's bedroom, even for a night at one-third the rate of prefabricated hostelries elsewhere that are paying some name-brand chain a percentage of their gross, just for the right to fly the corporation's colors in the lobby that was designed by a computer somewhere in a suburb of Dallas.

The Bellmard was designed on-site by the same carpenters who built it back in 1881 when stagecoaches and horse teams still plied the road out front. Some of the twelve rooms are a little smaller than others, according to the planks available back then. The wooden floors are well worn now, so a few of the tougher knots stand out a little beneath a bare foot. It has been a tourist home since the beginning, with the latest proprietors being the Butlers (he's from Ohio, but she is a Maine native).

He was supervising an auto-body shop in central Ohio. She was wishing she were back home. She saw an ad offering a country tourist home for sale in Princeton, Maine. It wasn't on the coast, but it was all they could afford. And the sellers said it was a going business.

Of course, it wasn't. The Butlers had to struggle, packing people in so tightly in the busy summer months of the late 1960s that their three children had to sleep on the couches some nights. Not during the winter, of course, unless a blizzard closed the highway for a day or so. The mealtimes were real homey with everyone chatting while they passed around the antique platters of fresh foods, especially the homemade breads and doughnuts, which were Letha's speciality. "They went over quite well," Letha recalls. The Butlers met lots of interesting people from places they had never been. Letha even waited on Governor Nelson Rockefeller one evening. Savoring the memories of those pleasant hours got Letha through all the mind-numbing sessions of bed-making upstairs, which is better than picking potatoes as a nine-year-old for four cents a barrel. And the Butler children helped with the beds.

But somehow sometime on a date not noted in their old guest book, things began to change. To be sure, prices did rise, all the way to fifteen dollars a person. Clientele, like bags of grass seed, always includes a certain proportion of weeds, the kind of guests you'd like to help pack. But the weed seeds were growing. Fewer families came. The couples usually came without their lone child. The proud parents, who are grabbing a few days away because with the hectic pace of their two jobs, you know, there isn't much together time, can still produce color photos of their youngster, who typically stays with an elderly neighbor for a reasonable rate, given the price of baby-sitters these days when grandparents live so far away.

Of course, a few of the traditional couples do bring their child, although even those families, it seems, can't afford to linger long. They're always in a hurry to get somewhere else, even on old U.S. 1. And the youngsters don't seem as well-behaved as the last generation. But maybe that's just Letha's memory; she's in her sixties now, you know. But, wait, the parents aren't as well-behaved, either. Some knickknacks seem to disappear from the bedrooms now and then. On occasion, even the large bar of soap from the bathroom down the hall seems to slip into someone's bag.

The most expensive change, and, therefore, the worst, was the meals, which became leftovers before being served the first time. People would say they were coming for dinner, all nice and eagerlike. But then they'd be late. And when the Butlers saw them coming up the walk and began getting the warming food back out of the oven, the people would say, oh, no, thanks anyway, they had been struck by hunger an hour before and, since there wasn't a McDonald's on every corner way out here, they had dined at the new pizza place just across the way or at one of the hot-dog stands that sprout along the sides of U.S. 1 come summer. And, by the way, where was the TV? They did have one, didn't they?

The conversations, thus, were fewer and less interesting. The guests were really up on all the TV shows and all the instant stars that dot the screen so briefly, but they hadn't read many books, probably due to how busy their lives were. Willis, who had taken a day job at the large Georgia Pacific paper plant to make ends meet, felt the younger guests seemed to believe the world owed them a living. Not his three children, naturally; they had all gone off to find jobs and found families elsewhere, though they did stay in Maine. So the Butlers can see their six grandchildren regularly.

Sport fishermen are good guests. So are the elderly, who dodge the younger crowds by planning their trips after school reopens when the leaves are pretty. And a fair number of Canadians still come over for the night. They're okay, too. It's funny how, originally, the area's white residents would flee from any Indian in sight because the Passamaquoddy tribe was not pleased with its ancestral hunting grounds turning into tidy farm fields and messy barnyards. Now the Indians advertise on the radio and the whites flock to the reservation for the high-stakes bingo games that can see some winners walk out with five thousand dollars or drive out in a new car.

Except for Indian bingo, the local clubs are dying off so fast you can't count 'em. Even the snowmobiling clubs, where members spend an evening riding from house to house to consume a different course of dinner at each, are melting away. It's warmer just to stay in the basement. And the younger people, who like all that outdoor sort of thing, are moving away to find work. Judging from all the local talk, the ones who stay behind are not so happy, either. Some drugs have crept in; so has fear of crime.

So now, the Butlers, only the third owners in more than a century, have become the first to keep the door of the Bellmard Tourist Home locked at night against strangers.

Ammo Across the Border

TEN MILES DOWN THE road right across the St. Croix River from St. Stephen, New Brunswick, sits another changing community. Nearly two centuries ago, Massachusetts, in appreciation for French help during the Revolution, thought it would be an honor to name one of its isolated outposts after the French port of Calais (as in ballet), especially since the Maine community confronts Dover Hill over in New Brunswick. And if those Massachusetts Francophiles didn't know how to pronounce those foreign words quite right around Calais (as in Dallas), at least the friendly intent was there.

It is a tradition that has lived on, thanks to the physical realities of love and the mutual isolation of two communities in two nations that have more in common with each other than they do with their own countrymen (and women). These feelings spanned more than the St. Croix River. During the War of 1812, the folks in Calais ran so low on ammunition that when July Fourth came around, they had no explosives left to celebrate. So the people of St. Stephen, a short cannon shot away, loaned their neighbors a goodly supply of gunpowder.

That kind of practical friendship permeates the re-lationship along much of the Forty-ninth Parallel, where the only things that confront each other across the frontier are the flapping flags, the matching church steeples, and the local softball teams come summer. Back up 1, the communities of Madawaska, Maine, and Edmundston, New Brunswick, found a way to get around the tariffs on wood and share some vital jobs. They turn the wood into pulp on the Canadian side of the bridge; the politicians, you see, forgot to mention pulp in their regulations. Then the Canadians pump the pulp through a pipe beneath the river bridge over to the American side, where they turn the slush into paper. No wood. No duty. No problem.

When the area was first settled nearly two centuries ago, no one was sure in which country the area was. It didn't matter much since not much seemed to be going on there. Today, countless families are divided by the border but tied by blood, which is thicker than water even at northern Maine's winter temperatures. They are united by spouses with jobs in different countries or Social Security checks coming from both lands or sons and daughters and in-laws across the river. When a new Canadian customs officer in one small town tried to tax the fresh turkey that some Americans were taking to their Canadian church's dinner for Canadian Thanksgiving in October, he quickly found himself transferred back to the city, where that kind of thinking is acceptable.

Without thinking and without any border formalities at mid-bridge, the Calais and St. Stephen fire departments each respond to the other's calls. The city of Calais even buys its water from Canada, and then resells some of it to nearby Milltown for a welcome profit, with the pipeline running right alongside U.S. 1. In between the want ads on the border's local radio stations, announcers must give the time in both countries. ("It's nearly one o'clock in Calais, nearly two Atlantic time.") But even the time difference is turned to a benefit when, for instance, the nightly closing of Canadian bars leaves one more hour to imbibe across the bridge; this sixty-minute gap also permits two New Year's celebrations, although it gives rise to some confusion on the exact birth date when a new child arrives around midnight.

In the early days of this century, virtually everyone in both communities was born Canadian. That's because the only hospital was in Canada. So at age twenty-one, everyone would have to pick a country. Like as not, the husband would pick one country and the wife the other just to keep their options open should there be another revolution or a change in Medicare.

Sammy Sanders's parents were raised in Canada in an area of St. Stephen known as Skedaddlers Ridge because that's where Maine's Tories fled during the Revolution. Sammy's father was a carpenter; he built his own house, in fact, and it's still standing, unlike some of the newer ones.

The five Sanders did not live with the rich folks up on Hinckley Hill, although in a town with an official altitude of nineteen feet, even a mere rise becomes a hill. Instead, the Sanders family lived in Calais over on Mud Lane, now called The Union. It was a tough blue-collar Irish neighborhood dominated by the Immaculate Conception Church and the Donovan, Davidson, and Driscoll clans. Mud Lane was so tough, it was said that the devil needed extra time to fetch his recruits from there; they didn't go anywhere without a fight. But even though the Sanders were Baptists, they were accepted on Mud Lane (not by accident, Mr. Sanders kept a more important secret, his family's Democratic party political allegiance).

Mud Lane was appropriately named. Roads were informal affairs in those days. A precarious stagecoach path ran the ninety-three miles from Bangor to Calais, eighteen hours of wilderness and fear with six changes of horses. It was another thirty-six hours by stage from there down to Portland. But well before World War I, records indicate, Maine contained 77 motorcycles, 715 automobiles, 989 licensed drivers, and 22 car dealers, a proportion of about 45 drivers per car dealer, which seems to hold true even today on many urban sections of U.S. 1. When there were only two automobiles in all of Bangor, they both collided at the corner of Main Street and the Presque Isle Road, creating Aroostook County's first car accident. At that time, highway patrolmen did only road-maintenance work, and they had to provide their own horse to get about.

To go anywhere else from Calais, folks took the coastal steamer, unless it was merely a winter Sunday's outing by sleigh on the frozen dirt of what would become U.S. 1. And for a while, Herb Hanson ran a stage through the mud down to Eastport. But there didn't seem much call to go anywhere else. Family was here. And so was work. There was shipbuilding and the cotton mill and the shoe factory, not to mention the Ganong candy factory on the Canadian side, where some genius invented the chocolate bar.

Slowly, gravel worked its way up the former paths of the Maine seacoast, easing the uncertain voyages of the Model T's and A's that were appearing more. Still, anyone planning a trip needed to ask locally about the condition of every segment ahead. Road signs consisted of occasional colored bands wrapped around roadside trees, and many Americans, farmers included, railed against the noisy, unpredictable cars. They frightened the main mode of transportation, the old reliable horse, which was still needed to pull those newfangled toys from the mud. Leagues of car opponents were formed. One suggested requirement: Every car must have a flagman walking ahead to alert others on the road, and each driver must stop at least once a mile to fire off a warning rocket.

Slowly, town-to-town jitneys began to emerge, along the coast, anyway; they earned extra income by mysteriously but predictably developing the need to have a rest stop and deliver their captive customers to those local merchants who had made advanced "donations." By the time Sammy and the other local boys went away to work for the CCC in the Depression, U.S. 1 was still unpaved. But they could ride to the government camps in the mayor's own truck, which seemed huge to the young men because it stood tall enough to clear the ridges of dried mud left behind by previous vehicles as mute testimony to their dirty struggles.

The old roads wandered back and forth all over the countryside. They didn't design the roads to have scenic turns that seemed to caress the countryside. In fact, they didn't design the narrow roads period. In those days, there was more time to spend going around obstacles than there was mammoth machinery to use conquering rushing rivers, erasing minor forests, or humbling obstructive hillsides. So, they never thought about building roads that ran straight as a shot from point A to point B. That would come and be called progress.

Early supporters of automobiles (and the ensuing proponents of better roads and then rural mail delivery and then rural electrification and then even instructional satellite TV) hailed their invention. They said it was yet another advance that would enable and encourage the nation's scattered rural masses to stay in the countryside, the treasured reservoir of Jeffersonian values, and thus avoid further populating the polluted cities, those compacted collections of corruption. Oh, sure, better roads—it might be more accurate to call them less worse—helped more folks come to town for a Saturday. But those were to be merely outings. Some excited day-trippers parked their cars on Calais's Main Street Saturday morning to assure a prime spot to watch that evening's parade of socializers, many munching roasted peanuts from Bernardini's. Before the age of MTV, there were Sunday-afternoon band concerts, unchaperoned excuses to shepherd a sweet-someone somewhere, perhaps capped off by a hot chocolate at Janet Todd's confectionery.

But the roads that carried people into town could also carry them away, far away, in a dynamic society that always seemed to be plotting changes just a few paces faster than the institutions and individuals trying to keep up. Calais was not the first farm town or mill town to lose a factory. The first industries, of course, were lumber and shipbuilding. The launching of Calais's first ship, the *Liberty,* in 1801 would also launch a business that lasted for decades. Calais's next industry was shoes. Shipbuilding lasted only as long as those vessels needed tall, strong masts. Shoes held on a little longer. But like the thick old elms that no longer line the Main Street of Calais and countless other communities, the manufacturing industries died off.

First, following the forces of economic gravity, these vital employers went to the larger American cities that seemed to suck people and opportunities into their crowded, dirty streets and to change their ways, something the parents and domestic emigrants really didn't notice until the offspring happened home on visits. Later, of course, the modern magnets for manufacturing jobs became foreign cities that drew in their own former farm folks to the ill-lit assembly lines and, understandably, left the former American employees to find other ways of survival in a harsh new world of economic Darwinism. It's the move to what is called a service-oriented society. People don't make things anymore; they perform services, from flipping burgers to maybe fixing the products made overseas. And nowhere in America would these changes be more obvious than all along the former stagecoach road called U.S. 1.

They drifted away from home down 1. The Butlers'

children left their parents' tourist home in Princeton, Maine, to settle in Ellsworth and Auburn; he manages a paint store and she married an ex-sailor who helps build ships in Bath. Sammy Sanders's brother and sister sought work down in Freeport and Belfast. Sammy always felt more at home in Calais, so he stayed. He was the bookkeeper at the Lincoln dealership and then he was business manager at the hospital. And then after the Eisenhower era, Sammy Sanders was appointed Calais postmaster by President Kennedy, an honor (and reliable job) that even Sammy attributed to the fact that he was just about the only living Democrat that even a president from Massachusetts could find in that part of Maine.

Sammy worked hard at postmastering. He was perhaps the most successful Second Baptist in town, or ever thereabouts. Friendly. Open. Funny. So successful was Sammy that he was not unappointed when the Republicans took back the White House for most of the next generation. Much of his success seemed to be due to his gentle demeanor, which stemmed from his affinity for dogs and pigeons. "It doesn't seem strange to me," he'd say. "I've got pigeon fever ever since childhood." Pigeons are not famed as an aggressive species, nor are collies. Sammy raised them in handmade shacks out back, where they would set up a racket of friendly barks and welcoming coos whenever the diminutive Democrat neared. Sammy would retire to the sheds after long workdays that doubled in length at tax time to handle his sideline accounting business. Sometimes he'd make little wooden sleds to sell at Christmastime or travel as far away as New York to swap stories and pigeons with other fanciers and, a few times, to win trophies for his birds.

One time, the postmaster of Calais was fined fifty dollars for forgetting to declare some birds he was transporting from Canada. The judge was unswayed by Sammy's argument that the birds fly back and forth across the border every day, equally undeclared. The next weekend, the same group of birds found themselves in tiny cages hanging inside the engine compartment of Sammy Sanders's auto, moving with uncharacteristic deliberation across the bridge from Canada and up U.S. 1 before turning on to Harrison Street after passing Clark's Variety Store.

But that time, unlike almost every other day of his life, Sammy did not pull into Clark's graveled parking lot. You see, after his retirement in 1982, Sammy became an Over-Homer. Over-Homers are ten regular lunch-munchers at Clark's sandwich counter, all retired men. They dine on self-appointed stools (Sammy's is second from the near end) and debate every conceivable social issue before buying their state lottery tickets, together for good luck, and then returning "over-home."

"Every day," said Paula Clark, a clerk and niece of Eleanor Clark, the owner, "they solve all the world's problems." All the problems save how Paula can escape Calais, since the navy told her she'd have to wait a year and the Border Patrol seemed slow to respond, too. "It's too dead in Calais," said the nineteen-year-old, "nothin' to do 'cept clerking." So she womans the cash register in the store that sells everything from sodas to cheap toys, sewing magazines to rifle ammunition, from snuff to video rentals every day until 10:00 P.M. (U.S. time) from 7:00 or 8:00 A.M., depending on when Eleanor gets around to opening. To speed checkout, every item carries two price tags, one for American money, which is green, the other for Canadian, which is all colors, has a woman on the two-dollar bills, and a loon on the silver dollars, which are gold.

"I've gotta get outta here," said Paula.

Of course, not everyone wants to leave. And others would like to see some come. Therein lies a philosophical and economic dispute that is brewing along much of U.S. 1, and elsewhere. It pits those who oppose growth because it threatens what they like best about where they stayed against those compromisers who also like where they are but favor at least some growth. They see any changes as an unfortunate but necessary price for the jobs that more prosperity brings. The compromise that seems to be emerging in Calais and to the south is to seek more tourists, who bring in dollars and take away Indian moccasins from Taiwan.

The foundation of modern tourism is that many of those people who ran away to their future in the cities, and their grown children who have known only suburbs, now feel the annual need to flee the pressures of those pricey places and seek solace in the refreshing sights, the natural sounds, and, above all, the exhilarating elbow-room of the countryside, even if it isn't exactly the same countryside their family left years ago. Tourism creates local jobs and, all right, maybe some traffic congestion at peak times and the visitors inadvertently sit at a regular's table over at the café. But tourists don't create impossible pollution, unless you count plaid shorts.

Of course, like migrating geese, some of the outsiders end up staying and fouling up the place. And while this leads inevitably to the appearance on local grocery shelves of some totally unnecessary luxury food items, even foreign wines, the arrival of new blood and, more importantly, new money, can help firm the sinking real estate market. The surviving local contractors don't mind putting in some new foundations and septic tanks. And a lot of these newcomers are suddenly converted to conservation, the definition of a conservationist being someone who built their summer home last year.

The people who live in these areas like them or else they'd have gone somewhere else. Good riddance. And the people who don't live in these areas don't matter. It can be a difficult adjustment personally and community-wide to be told that after a century or two of isolated local life, they now must begin touting themselves to a bunch of strangers, albeit affluent ones, with out-of-state plates.

The trick in communities unaccustomed to such hard sells and boasting about home is finding and then marketing The Hook, the irresistible attraction that will draw so many wallets so far. It is no longer sufficient to be plainly rural; even the Pennsylvania Dutch have gone commercial in the souvenir shops along the Interstates through those areas.

Once, it was sufficient for an area to be simply scenic. That still helps, since so many areas aren't anymore. But scenic alone won't hack it, according to all the economic consultants with their Attraction Analysis Reports. There must be something to do, too. An old fort is a start, maybe with some locals dressed in period costumes come summer. There should be some shops stressing quaint things; small wooden barrels of scented candles are popular. So, too, are 'n's, as in all the stores with names like Buttons 'n' Bows and Books 'n' Things. And a water slide can keep the kids busy.

Calais and St. Stephen have little of these—yet. What they do have are a core of younger people, mainly newcomers of varying degrees, with some ideas. They include Susan Harvey, who moved up from Bangor a few years ago to put a little zip into the Chamber of Commerce, and Billy MacCready, who is the energetic former mayor of St. Stephen. Mrs. MacCready is a less-new newcomer, coming from the can-do province of Alberta in western Canada. She was the founder in St. Stephen of Mrs. Billy's Restaurant, Home of the Happy Cooker, which enjoyed short-lived though satisfying success.

With the kind of open-faced pragmatism that comes from being unaware of historical inhibitions, they suggested a new kind of cooperation among Calais and St. Stephen and twenty-five other adjoining communities. The Americans still needed jobs to stem the loss of their young, and the Canadians did, too, although the local Star Kist tuna processing plant was a help. The two women set out to create new opportunities for economic expansion without going too commercial. "We don't want to become another Bar Harbor or Freeport," said Mrs. Harvey. This was not an imminent threat way up there nearly 800 miles above Washington and 550 from New York. "But we've got to do something." Most any light industry would be welcome, although the area is a ways from major markets.

Fishing for the bounteous waves of alewives that once populated local waters didn't seem likely to hack it as a major attraction, either. So the heart of their plan centered on the Quoddy Loop Project, a scenic looping drive back and forth across the border, through the two communities, and down U.S. 1 along the shore to Quoddy Head State Park, easternmost tip of the United States. By bridge and ferry, tourists (or anyone, actually) can tour three islands, two of them Canadian, including once-famed Campobello—now home to an international park—where President Roosevelt liked to live (and caught polio after fighting a forest fire in 1921). Pooling their resources across the border made a lot of common sense, which rarely seems to guarantee anything these days other than a lot of trouble. At one point, seeking government subsidies, the pair shepherded Canadian officials around the entire loop. The women and their committees had arranged donated cars from dealers on both sides, the same for ferry operators. They even got a U.S. Coast Guard vessel to meet them at one wharf. The visiting officials from Ottawa seemed shocked at the scarcity of red tape. Privately, they quizzed Billy on why her office was actually located in the American chamber's building on the U.S. side. "Simpler and cheaper," she replied, which didn't seem to make sense to them.

Nobody believes these ideas will work without getting them on TV. Billy even wandered down to some Eastern Seaboard cities the old-fashioned way, on a coastal steamer, seeking to create photogenic publicity for tourists. But the best publicity idea was to launch a summer festival, emphasizing the international cooperation of two separate communities. It was probably Billy who came up with it during one meeting. "Remember how St. Ste-

phen loaned the Americans some gunpowder during the War of 1812 to celebrate July Fourth?" she asked the group. "Even while we were technically at war? Well, why don't you Yanks return it?"

That prompted several heads to nod. "Well," said one of the Americans, tongue firmly planted in his cheek, "how much did you give us?"

"Yeah," said another, "and do you have an invoice to prove it?"

The idea caught hold, though. They got an antique wooden keg. They attached a commemorative brass plaque. For a note of authenticity, they filled the keg with real black powder, though some thought it looked suspiciously like black pepper. And they scheduled the payback ceremony for August, at mid-bridge.

To everyone's surprise, however, the display of abiding goodwill did not attract much media attention. The TV producers asked what was going to happen at mid-bridge, and when Billy described the nice little ceremony, the producers' interest immediately waned. "They said it would be much better if there was an explosion of some kind," Billy recalled. "They said they needed tension or confrontation to film. I tried to explain that the whole point was international goodwill."

But the cameras did not come that day. Friction, yes. Friendship, no.

(Shortly after we met, Sammy Sanders went "over-home" for the last time. But it wasn't the devil who fetched him.)

Halfway From Up There to Down There Is Here

U.S. 1 OUT OF CALAIS is no longer lined with the big old elm trees that lived so long before succumbing to the devastating disease that arrived in the early part of the

century on a lumber ship from the Netherlands. But just south of town, the birches line the road like white ghosts. A hardy breed that spends nine months waiting to grow for three, some of them are as thick as a man's arm. All are bare, of course, their massive numbers sprinkled with evergreens stuttering through their own growth cycles.

Both species surround a small roadside park a few miles south of town. There, a couple of frigid barbecues holding last year's ashes peer above snowdrifts that have shut the log-lined driveway. Pale green moss, the kind beloved by deer come winter, hangs from numerous branches. In places, the snow seems green from the millions of needles shed in this year's batch by their nearby parent trees to begin the long decay process back to dirt to feed and house the roots of the host trees.

Sitting in the snow like a grave marker is a reddish granite block with a metal plaque: THIS STONE MARKS LATITUDE 45 DEGREES NORTH HALFWAY FROM THE EQUATOR TO THE POLE—1896. It still is halfway. So is the Rest Lawn Cemetery across the highway.

U.S. 1 all along the 3,500-mile Maine coast is a winding, sometimes tortuous byway as it follows some broad point out toward the immense sea and then ducks back toward the safe shore, time and again. It gives off a very strong sense of driving back through time to another era when straight lines and making time were not quite so important, when the goal was not necessarily to get from here to there as quickly as possible but to enjoy the getting instead of just the arriving and having. The passing roadside is frequented by many old houses and huts, sheds and trailers. At some, a rusting relic, a veteran of the adjacent highway, rests without wheels in the yard, along with a once-orange basketball hoop awaiting another season. Or a blue Ford tractor, its small size revealing of the tasks it is asked to perform on these modest farms, sits tucked hard by an old barn. The crooked roof still keeps the heavy snows away but not the cold, which creeps in everywhere. Many bare trees flock about like haunted stick figures, but the power poles are the thickest things. Countless woodpiles stand stacked and dry beneath the blue plastic tarpaulins that have replaced the faded old green canvas coverings in the modern countryside. The sweet-smelling smoke that drifts slowly from nearby chimneys foretells the logs' future fate. "The homes," according to one writer in the 1930s, "show few evidences of prosperity, being weather-bitten and unpainted." They still are.

Half-frozen dirt roads run off to the side occasionally toward the deep woods that still yield the wood for lumber, pulp for paper, many thousands of boughs for Christmas wreaths come fall, and, in summer, many millions of pounds of berries, which are delivered to all the Berry Receiving Stations that mark the roadside. On winter days, loggers park their muddy pickup trucks at the dirt road's junction with 1 to ride larger rigs to the work site where chunks of bark, mounds of chain-sawn chips, and broken limbs tell of the harvest that has happened. Other lanes wander down toward the sea and a few sparsely populated islands where fishing and oral legends once abounded because without television and the telephone people socialized in person and passed on stories they had heard, possibly with embellishment in a subdued Down East kind of way, of course. There was, for instance, Barney Beal. He was dubbed Tall Barney because he was six foot seven and disliked disrespect so much that he broke a few arms, even if they belonged to British authorities shooing his boat away from Canadian waters. Sitting in a chair, which he sometimes did, Barney's arms could touch the floor, which means that either it was a very low chair or Barney would have been a first-round choice in the NFL draft.

When the powerful eighteen-plus-foot tides go out on their chilled winter cycles, the many streams ducking beneath U.S. 1 seem to go dry. And the brown-bottomed coves and bays are littered with glistening rocks and off-white chunks of sea ice the size of tables and cars. It was the southerly migration of grinding glaciers from the Arctic that gave Washington County its sometimes still-scarred look. (Those visitors from other climes who attempt to go swimming may suspect the glaciers still remain just around the point.) And, a local historical pamphlet notes, "Residents are what they are because of where they are." So beware.

In Eastport, the United States's easternmost city, which the British held for four years during and after the War of 1812, old-time merchants once harnessed these twice-daily watery shifts into tidal-powered gristmills. Even though this is overalls country, eight tawny deer gently forage in an open field not fifty feet from passing cars, seemingly aware that deer-hunting season is over.

In the distance, children skate haltingly across frozen roadside bogs. Many fruit stands sit bleakly along U.S. 1 here waiting for next summer's crop of blueberries, strawberries, and tourists. Berry picking and packing have actually become substantial industries as shipbuilding and even lumbering have waned. The berries, carried by birds and bunnies gorging and then relieving themselves in the newly cleared openings, spread rapidly.

Along the road, fading signs, tilted from the snowplows' passing blasts, tell of summertime activities, advertising helicopter rides and the Blue Beary Snack Bar (CLOSED FOR WINTER). For January drivers, the country roadway is punctuated by the rhythmic *ker-thumk-ker-thumk* of tires hitting the buckled pavement's frost heaves every twenty or thirty feet. They, too, will melt away come May.

Machias (muh-CHY-us) and all the little Machiases around about are, like many other coastal outports, full of history as fishing ports, lumber ports, and pirate ports. Over a century before the Revolution, notorious brigands such as Captain Rhodes and Samuel Bellamy operated out of Machias (in Indian, the name means "bad little falls"), helping to relieve overloaded seagoing ships of their treasure. During the Revolutionary era, allegiance switched to the emerging homegrown government, which did not oppose smuggling if it hurt the British and which didn't have much military clout anyway to interfere with business.

In those parts, in those pre-911 days, the law was whatever was said by the nearest person with the largest weapon. At times, the pirates operated with the unofficial sanction of the government. This was called privateering, which was politically useful for governments and might at times bring in some helpful ill-gotten gains, and, making the stealing once-removed, helped combat governmental guilt. Privateering was also excellent training for pirating, which is what privateering became when the seamen began pillaging the ships of their sponsoring government. Sometimes governments would dress sailors up (or down, depending on your fashion standards) as pirates, which enabled them to unofficially act like brigands without further hurting their own official reputations.

Soldiers wearing the outfits of another profession may have offended the pirates' sensibilities. But pirates weren't always so pure, either. Sometimes pirates would fly the flag of whatever government struck their captain's fancy that day, his fancy often having a lot to do with the flag being flown by the nearest ripe ship. This could, however, result in two boatloads of pirates pretending to be the same nationality so they could get close enough to attack each other, which was okay if one group was privateers. But pirates attacking pirates could be professionally embarrassing, so the losers had to be killed.

Along the isolated Maine coast, none of this was necessarily deemed illegal, at least not by anyone in a position to do anything about it. And Delaware had not yet been designated as official corporate registration point for domestic ships.

Because all this could be quite confusing at times, the famous Cotton Mather, preaching farther down the coast road, made pirating the subject of one of his "hanging sermons" in 1704, which didn't end seafaring thievery, because a lot of the pirates did not make it to church that Sunday. Nonetheless, Captain Bellamy continued his wayward ways, fortifying Machias and portraying himself in the media as a well-intentioned Robin Hood–type character, although Errol Flynn had not yet been born. Like the future Flynn, Bellamy's position was usually proclaimed to captured crews from the forecastle of his ship, the *Whidaw*. Bellamy's political platform, which stressed the bottom line and did not mention health benefits, argued that the sailors had as much right to steal as did their shipowners, who, the captain said, were just more powerful bandits who had acquired the protection of politicians. To the new captives, joining Bellamy's merry band seemed definitely do-able, especially since the alternative was called walking the plank. And if they got to wear those nifty long-sleeved pirate T-shirts with the thick colored stripes around their chests, then that was a bonus.

So they signed up—or seemed to, anyway. One day, the captain of a captured whaler out of New Bedford, Massachusetts, joined up and offered to help navigate Bellamy's ships through some coastal reefs, which he knew well because he followed the whales through there. That is a Maine summer migration that those huge mammals still make. The whaler captain may have been among those sitting on the hard benches listening to Brother Mather. He intentionally led Bellamy's ships on to the reefs. Just about everybody drowned because, like the *Titanic* not too far to the northeast but far into the future, they didn't have enough lifeboats.

On the brighter side, however, such shipwrecks were good for the business of the numerous salvage companies that formed on the mainland to tidy up the shoreline and put such wrecks and their recovered cargoes to profitable use.

Off Machias on June 12, 1775, some local firebrands displayed their annoyance at the thought of supplying ships to the British and lumber for the barracks in Boston, where London was sending more troops to quell the futile guerrilla rebellion. The rising spirits of independence were augmented by the liquid spirits quaffed during long hours of plotting at Burnham's Tavern on High Street, which later became U.S. 1. The plot was to take on and capture the armed British schooner *Margaretta,* which they did with some local boats. This was yet another of the naval battles proclaimed as the first in the Revolution. Its success was due in large part to the little-known heroism of Hannah Weston, who, while the boys were plotting down at the tavern, collected fifty pounds of powder and lead from homes as far away as Jonesboro and lugged them seven miles or so through the woods along the coastal path.

A month after the *Margaretta* incident (things didn't need to happen so fast in those days), the British retaliated with a bombardment by four ships. The battles themselves didn't matter a whit, except that as initial skirmishes often do, they provided Revolutionary orators with ample ammunition to justify acquiring additional arms, in this case, a Continental navy. (But more on that when we get farther south on old 1.)

Today, Sergeant First Class Rogers at the Machias Army Recruiting Station uses more modern methods to boost enrollment: a telephone answering machine, which takes the phone numbers of would-be recruits and reminds them, "Don't forget to ask how you can get up to twenty-five thousand dollars toward a college education."

Many residents of coastal Maine, however, are too old to enlist. Some like Sarah Smith can stay at home with their relatives and memories. Others can't, or don't. And their relatives deposit them in one of the many nursing homes that seem to be emerging quietly in towns across the country to meet the social and commercial demand for custodial and health care once routinely provided at home by families.

About an hour's drive down 1 from Machias is Ellsworth, which bills itself as "The Crossroads of Down East Maine" and is named for Chief Justice Oliver Ellsworth, who never saw the place. Near Ellsworth, in a setting that still seems placid and rural to passing eyes, is the Maple Crest Nursing Home. Originally a country inn, Maple Crest now has thirty-four residents, most of them from the area. Each pays about eighty-five dollars a day, and Joyce Jamison, the registered nurse, estimates that perhaps a dozen of them can still feed themselves and know who they are.

The nurse and other workers can almost guess each family's history from the pattern of their visits. The close families seem to visit a lot, and several family members will come on their own at times spread throughout the long weeks, even when their loved one becomes less aware of who is visiting or what day it is. They will chat with the patient and eventually at him or her. Some no doubt believe there remains unconscious recognition or unspoken comfort, even for those where senility has taken over and alertness has completely vanished. There will be regular flowers and likely some children's drawings for decorations.

The other types will initially come in a group to register the elderly relative. They may visit in a group, too, for the first few times. But quickly the visits taper off; their lives outside are so busy, you know. And the bouquets come twice a year, probably around Christmas and a day after the patient's birthday. As Ms. Jamison and the government's statistics point out, medicine and technology are enabling people to defeat many once-insufferable diseases and conditions now so they can suffer and die later from the next level of unconquered diseases and conditions.

In Ellsworth, which is the commercial capital for a resort-rich Hancock County that encompasses Bar Harbor just offshore some forty miles from highway 1, Sarah Lewis still lives in her family's old home over on Pine Street. She well remembers when times were different, when man (no doctors were women in those days and no nurses were men) could do only so much when sickness struck. She lost her father eighty years ago. "It was pneumonia," she recalls. "You couldn't do a thing about it in those days."

He was fifty-four. She was twenty-one. That means that Sarah (she always hated the nickname Sadie and is waiting for the last person who remembers it to pass away) marked her 101st birthday not long ago. "I've had a wonderful life, interested in lots of things. I played the piano for the church and some little ol' dances and for the silent movies, too." The movies were exciting at first but quickly grew boring time after time after time. So she turned to watch the people watching the movie. And there was always a singer at the intermission.

Ellsworth was another port whose fortunes were nailed to wood, with more than 150 nineteenth-century schooners working their way up and down the Union River with their loads of wood from the wharves. At one point following the great potato famine in Ireland, hordes of Irish immigrants moved into New England and Maine, where prohibition was popular before Prohibition. Ellsworth and the *Ellsworth American*—an outspoken proponent of the American, or Know-Nothing, party, which favored government only by native-born Americans— became synonomous with racial intolerance. Father John Bapst, a Jesuit missionary adrift in a land of rabid Protestantism, sought to exempt his group of Catholic students from having to read the King James version of the Bible. Such an outrageous idea prompted his immediate transfer all the way to Bangor and approval at a special town meeting of a resolution vowing that if the priest returned, Ellsworth's citizens would manifest their gratitude "by procuring for him and trying on an entire new suit of clothes such as cannot be found at the shop of any tailor."

Sarah Lewis's father was from a seafaring family and, not surprising in such a town, became a carpenter; in fact, he built the house his granddaughter now owns. It was not among the 130 structures consumed in the minor stable arson that quickly became the locally famous fire of May 7, 1933. However, the fire did get the home where Sarah lived with her husband, Lewis, and their daughter (the other one died). But the family rebuilt the house. And Sarah, now a widow, still lives there. "I own my own property," she says, "though I don't go out much in winter."

Her husband was a rural mail carrier beginning right at the end of the horse-drawn mail sleigh era. He rode on U.S. 1 every day, even before it was numbered in 1925. It wasn't often easy; many days he'd get stuck and have to seek help from a farm not necessarily nearby. He couldn't phone home; there were no phones. She cooked on a wood-burning stove with an oil lamp nearby. "It was terrible," she recalls. "He was always late for dinner." But on some Sundays, they could use the car for a drive in the country, which began just a few blocks away. They'd pack a picnic. Sometimes they might cover as many as thirty-five miles. "Everybody loved the seashore in those days."

On January 2, 1900, Florence Woods, having obtained special permission, became the first female to drive a car in New York City's Central Park, telling a *New York Times* reporter covering the historic event that she was amazed at the temerity of other American women. Such urban audacity took a while to seep into the Ellsworths of America, so Sarah let her husband do the vehicle's manhandling on those Sunday drives in the early 1900s, back before ever so socially conscious newspaper editors felt compelled to change such a word to "personhandling." Sarah stopped driving in the late fifties because of her eyes. But she goes on rides in good weather and senses some disturbing changes. "You wouldn't know the town today," she says. "It's a trouble."

First of all, there are so many more people because so many have had to leave the countryside to make a decent living. In and around town, there's so much building going on everywhere—houses and lots of rows of little businesses—and that consumes valuable wildlands, which, she feels, upsets the natural balance of living things. But what do a 101-year-old lady and a bunch of rabbits and deer and wild songbirds know about progress? Such a natural balance can seem to matter even in developing regional centers like Ellsworth, where cement has not yet covered everything and some voters have seen local officials somewhere other than on a TV screen.

And there are all the newcomers moving in, too, looking for safety and clean air, affluent, mobile people who haven't spent a decade in the same suburb let alone a century on Pine Street. "We're getting a lot of foreigners here," Sarah says, "from outside the United States, too. Newcomers think Maine is not too friendly. But I think it's the other way around. Their manners and ways are just different. And since the outsiders don't like to change, they'd like us to adapt."

Sarah rarely hmmphs about her views to outsiders, who include one of the Butlers' children from the tourist home back up 1 in Princeton (the other Princeton is way down U.S. 1 in New Jersey and has its own problems along the highway). These newcomers have nudged Ellsworth's population from 3,557 in the mid-thirties to about 5,100 today, a veritable boom in the shriveling American countryside. Sarah simply shakes her head and goes on about her own business living and remembering from within her old house and, of course, attending church regularly. "I live right across the street from the Baptist church and next door to the Methodists. And I'm Unitarian, you see. So I need to be real good around here."

Betty Is Beautiful

A FREQUENT SIGHT ALONG U.S. 1 in the rural north is the simple tiny cabins that appear at the roadside end of driveways, often without even a door. They are, of course, basic winter wind shelters for youngsters awaiting their school bus in the unheated early-morning darkness.

The big yellow vehicles waddle their way up and down 1 according to a schedule reset each fall when the local schools figure out which families still remain to send how many kids to what grades. Every few hundred yards, the buses pause, their illuminated eyes alternating red winks side to side, and all the local traffic pauses, too, while their momentarily hot exhausts are whipped away by the winds. There are fewer youngsters in these districts now, a factor of both birth control and the faltering economics of rural employment. The bus drivers themselves have changed also. Many of them now are women, who are either single and need the money or are married and need the money. The hours are short. And driving a school bus is one formerly male-dominated job that the men don't seem to mind giving up. Some women drivers bring their own preschoolers along on their route to save on baby-sitting and those youngsters join in the confined chaos that moves along the road.

Time was when the male driver would have had just about enough of the raucous boys in the back, which seemed to happen more on the afternoon trips home after eighth-period geography. The men might pull their bus onto the road shoulder, set the brake, and march down the suddenly silent aisle to whip up a little pink silence back there. At those times, someone with twin pigtails was always eating an orange from a sandy Florida grove way down U.S. 1. The citrus stench mingled with the diesel fumes to ignite an occasional wave of nausea within some stomachs beneath the acrylic fiberfill of the colored K Mart jackets from tiny tropical countries where winter comes in June and means temperatures in the sixties. The women bus drivers still yell for silence at times, and it may work for a moment or two. But then the decibel count begins to mount once more. And the drivers let it pass, their modern training having included warnings on the potential for lawsuits following the administration of anything remotely perceived as corporal punishment.

Every morning on their way out of Ellsworth south up the hill past the Hale and Hamlin Law Offices, the gracious old homes, and the spreading apartment buildings that needed a controversial zoning variance, the buses grumble by beneath a handmade tree house. The impressive hideaway with the windows for peering down into passing cars perches between two far-reaching limbs of a sturdy old fellow that was a mere sapling when the first layer of pavement went in. In summer, the passing vehicles' invisible fumes rise through its shady mass. But only the lower layer of leaves turns yellow and shrivels. Some miles down the road, beyond the Ellsworth bus routes but before Bucksport, appears the first graffiti on southbound U.S. 1, as plainspoken as the rock it defaces: "Betty Is Beautiful."

For a while, as it has in several places over the years, U.S. 1 followed a different path out of Ellsworth. It ducked north for twenty-five miles to serve Bangor before following the west bank of the Penobscot River back down to the shore through Winterport (the height of river navigation come winter) and Frankfort, the 225-year-old community that began as a collection of log cabins where the British destroyed a couple dozen ships in 1779 and staged an abortive occupation in 1814. The river flats below Frankfort are among several reputed to contain buried treasure chests that once belonged to Captain Kidd, which explains the hopeful holes that seem to dig themselves overnight on the riverbanks from time to time. This guy Kidd gets around like Santa Claus and George Washington. From the number of places that know for sure he invested their riverbanks and beaches with treasure (CAPTAIN KIDD DUG HERE), he could not have had a whole lot of time left for pillaging. And Captain Kidd was not the last visitor to leave his money on the Maine coast.

Moving along the coast road, there is a strong sense of local traffic, with few vehicles traveling more than a few miles before turning off, some even using their turn signals. And there are, along U.S. 1, many, many, many places to turn off, most of them commercial establishments happy to see any traffic. At the Bucksport Dunkin' Donuts the other day, a large jar stood by the cash register to collect any loose change. It had been fueled with a five-dollar bill, merely as a suggestion, of course, and carried the label FOR MICHELLE. But who is, or was, Michelle? "Oh, that's me," chirped the waitress, who did not appear to need an organ transplant. And what was her problem? "I, uh, had a fire at my house and I thought someone might want to help."

Nearby is a once-abandoned gas station, one of many all along U.S. 1 that have been reborn to new commercial lives. This one now is a quick-stop liquor store that also sells sauerkraut, hunting licenses, and ammo for gunmen who have run short. (The Buckeye Taxidermy shop is just down the road in case a round or two hits something.) Some trace this openness about and abiding affection for violent weapons to the frontier minuteman philosophy, although the threat of foreign occupation has dimmed somewhat since three-cornered hats went out of fashion. Others blame the National Rifle Association. Either way, it amazes Canadians, who comprise the largest single group of tourists in Maine (upward of a third in many areas), as they do, too, at the other end of U.S. 1 in Florida.

Maine is the closest summer seacoast to Canada's population centers and Florida is the closest winter warmth. So Canadians, who must take a test like a driver's license to even buy a gun, marvel at how ubiquitous the weapons are among their seemingly civilized next-door neighbors. And amongst themselves, they tell the joke about the infuriated American who runs out to buy a gun to dispatch his landlord. "Well," says a suspicious gun dealer, "how do I know you won't shoot ducks with it?"

On Main Street in the town, one of many former stage stops along 1, is the Jed Prouty Inn (PRESIDENTS VAN BUREN, JACKSON, HARRISON, AND TYLER SLEPT HERE. ALSO ADMIRAL PEARY). Down Bucksport's main thoroughfare near the Verona Island Bridge also rests the grave of pioneer Jonathan Buck. For reasons not known by the living, his granite obelisk is marked with haunted graffiti, the shape of a woman's leg said to have been put there by a witch said to have been hanged by Buck for reasons left unsaid.

Heading out of Bucksport, the highway passes the imposing granite walls of the 150-year-old Fort Knox (not the moneyed one) before old 1, the Bangor Deviant, rejoins current 1 and wanders into Searsport.

Searsport is sometimes referred to as a major capital of antiques (the inanimate variety), a bid for dominance among visitors with credit cards, which no doubt would be disputed by upcoming parts of Virginia and North Carolina. Searsport carries on that grand American highway tradition of commercial form contributing to commercialism and rarely resembling art. The Lighthouse Gallery of Gifts is built in the shape of a lighthouse. Elsewhere on 1, there are buildings in the shape of milk bottles, tepees, and snowmen (but only down south where snow is exotic and not associated with shoveling). Despite their shapes, many of these establishments purvey that all-American staple of tourism: homemade fudge. The FTC, however, might want to investigate the homemadeness of fudge manufactured by the ton, wrapped in cellophane, and shipped from a loading dock in a new industrial park nowhere near any grandmother's kitchen.

Although Searsport's fame as a production center for expert seamen has faded with its fishing industry, a few independent lobstermen remain to harvest the unseen depths offshore where the crustaceans seem to be a lot smaller in recent years. And the port still handles varied bulk commodities overseen by Dennis Curtis.

As official marine surveyor of Searsport, he checks the bulk cargoes, certifying quality, quantity, and delivery as the referee between buyer and seller. Business deals need such checks for dishonesty now that they inevitably concern two parties who don't live in the same small town. Incoming in the large seagoing cargo ships that suddenly loom so huge on the local horizon of Penobscot Bay is fuel oil, tapioca flour for coating paper at the mills up the Bangor & Aroostook Railroad line, and tons of salt from South America and Indonesia, where it is mined or lifted from evaporating seawater in tidal basins. Some of the salt is tossed on the area's winter roads to melt dangerous ice and then seeps into overgrown ditches come spring, where it helps turn the grass brown and an occasional generation of field mice into mutants. Some of the imported chemical salts go to make sodium hydroxide for paper manufacturing. The paper, much of it bound for newspapers in the Middle East and Europe, is loaded back on the railroad and hauled down to Searsport's waterfront, where the brisk white caps whip right up to the pier pilings.

A fair number of Maine residents are what Newfoundlanders to the east call "C.F.A.'s," Come From Away's. They seem to like staying in and even owning bed 'n' breakfast spots, which can charge anywhere from $15 a night to $175 in season for a large room with its own bath and fireplace. Ross Overcash, for instance, left the frustrating profession of trying to teach science to today's college students, who already know they know so much because they saw it on TV. Ross runs a seaview bed 'n' breakfast with his wife, Mary Lou, in Lincolnview. This means trying to cover their overhead at the Red House with the income from two rooms rented nightly to middle- and upper-middle-class customers, including a growing number of women traveling alone who feel safer in such establishments.

Running a bed 'n' breakfast in Maine means taking reservations in May for October's elderly leaf-peepers. It means wan winter days of high heating bills, looking out on the lonely, empty highway. It means rising at 5:00 A.M. come summer for breakfast and bed-making. "You have to like people to do this," says Mary Lou, who does.

Some Maine towns seem to have more motel rooms than permanent residents, although Maine's coastal tourism accommodations are not forced to compete with those cookie-cutter communities that cluster around Interstate interchanges like travel tumors. Tourism, in fact, is about to become Maine's biggest industry, a trend that is new and good, since so many of the others are fading. But while uncut trees remain alive for future use, unused motel rooms are gone forever each night. This increasing interdependence of Americans and now even foreign countries, which began with the highways and continues with the jetways, does mean a larger potential market. But it also means less self-sufficiency in sales and supply. When times are tough in the corporate offices of Connecticut and Massachusetts, vacations are curbed or canceled, and times quickly get tough in the tourist homes, antique stores, seafood restaurants, and awfully quaint gift shops of Maine, where people quickly notice when out-of-state license plates grow scarce like caribou.

In Lincolnville, when Raymond Oxton was growing up just down the road, every family was basically self-sufficient. His mother cared for her own four children and washed clothes for a few other families. Everyone grew their own vegetables in the short summers. They picked their own berries for jellies and jams, and raised a few chickens for Sundays, the most dangerous day of the week to be poultry in the countryside. Milk came from a few family cows. Transportation was the family horse, maybe hitched to a buggy. Fuel was the hay or grain grown out back. Fertilizer was delivered from the rear ends of the livestock, which became beef when the animals got too old. And Grace Mahoney, the one-room-schoolhouse teacher who treated every student like her own child, could earn an extra twenty-five cents a week for stacking the school's woodpile and stoking its stove between the three Rs. Now the school's furnace, not to mention the power company's generators, uses oil that funnels from ships into big trucks at Searsport's marine terminal and rumbles down U.S. 1 from those faraway lands where men wear white sheets and women wear black ones.

In 1930, U.S. 1 was paved by hand. But before then, back when being pedestrian was not disparaging, you simply walked along the dirt road, which occasionally was sprayed with oil to keep the dust down. The next Model T that burped by would pick you up. And if, as often happened, the vehicle got stuck, you helped unstick it. To travel farther, one took a local jitney, a big old Cadillac or Buick sedan that packed every square inch of space for the trip to Bangor or Rockland, where Edna St. Vincent Millay, the poet, was born. In winter, back when Americans did not feel the need to tame seasons with snowmobiles or air conditioners, most people simply stayed indoors or attempted their own snowplowing. After a while, of course, the McLaughlin brothers got a pretty good plowing business going. Now, most everyone waits for the county trucks, since they already paid for them in their mounting property taxes.

Raymond, who has lived in the same house along 1 for all seventy-seven of his years, had a pretty good firewood business going until age, propane stoves, and electric space heaters sneaked up on him. He and Myrtle, his wife of fifty-two years, can't imagine what it has been like to live another way in the cities where keys are needed for everything. But he's starting to guess, unable to become "with it" and accustom himself to the TV shows filled with incredible violence, having heard recently that someone broke into a neighbor's empty house. They didn't take anything, which might have made sense. They just tore it apart. Every June, the sheriff's deputies can count on some overtime to process the rash of burglary reports as summer-home owners begin discovering what someone has done to their place over the winter. Raymond is also getting telephone sales calls from computers who don't hang up until they're done talking, from surveyors who want to know his brand of underwear and the last talking movie he saw, and from stubborn human salesmen who keep calling long distance about their long-distance services. "Whatever happened to dealing face-to-face?" he wonders out loud.

But not all the violence on TV is pretend. Maybe thirty miles from Raymond's safe home is the Maine State Prison at Thomaston, the first of several such intimidating institutions along U.S. 1—and more are coming, which is great if you want to become a guard. Along with tourism, prisons—or rather prison building and prison filling—are among America's fastest-growing industries as society's fears and impatience with real violence mount. Some experts estimate 1 million Americans will be incarcerated in state and Federal prisons soon. Some states have imprisoned more than 250 people for every 100,000 residents, the highest in the so-called free world. This is supposed to make everyone feel safer. Unfortunately, it hasn't affected the rates of violent crimes. So politicians and bureaucrats suggest the obvious answer: Build some more of the same prisons.

Four hundred and eighty of the nation's nearly 800,000 inmates live on the Thomaston site, which was sold to the state by Governor William King back in 1824 when no conflict was seen in such deals. Stephen Maxwell knows all the current residents. He's a corrections officer, the one who classifies them as they come and go, for education, living space, treatment, and danger.

Being classified as a corrections officer is itself a growth industry these days. While some urban areas resist new prisons or expansions of old ones, many others, typically troubled rural areas, compete to get them for their virtually recession-proof jobs behind the shiny coils of razor wire, for their $20-million to $30-million annual payrolls, and for their purchases of supplies, everything from pens and potatoes to masonry and movies rented for the prison's cable TV system. Maine is building a new maximum-security prison, its eighth prison facility, just down U.S. 1 in Warren. Its initial installation holds one hundred beds, but existing plans allow for an expected expansion to five hundred.

Like most American penal institutions, the nearly seventy-year-old Thomaston prison is overcrowded; administrators are trying to hold the population at 480, or 20 percent above capacity. So far behind the sentencing are the builders that many new prisons, costing upwards of $60,000 per cell, are way beyond capacity the day they open. Others must be expanded even during construction.

Part of this has to do with the drug business, dealing and stealing, which began emerging in the mid-seventies and was soon followed by a palpable public anger and frustration that focused on life terms that might last seven years. Politicians have not been adverse to mining those fears and angers. Concern heightened even more ten years later with the explosion of crack cocaine, an instantly addictive nugget of dope that is smoked.

So far, most of the political proposals and plans have focused on bricks and mortar. What goes on inside the prisons' perimeter within the less-visible counseling and educating programs has drawn less attention, except when it comes time to trim budgets.

Part of the growth also stems from new felonies; in some states, for instance, a third drunk-driving conviction now requires penitentiary time, as does the sentencing of habitual burglars, robbers, and so on, which often can make considerable sense when a case is considered individually, as it is by the juries and the public. It's only when the phenomenon is viewed by the unending busloads, as it is by prison administrators, that the flames of futility are fanned. And, too, the population of sex criminals has increased greatly with less social discomfort and stigma attached to reporting such cases. Thirty percent of Thomaston's prisoners are in for a sex crime; twelve

years ago, they totaled 2 percent. Although meaningful treatment of them is very difficult, officers say they do make model prisoners.

With little public pressure to cure or rehabilitate wrongdoers and great public pressure to warehouse them while at least holding down operating costs, few institutions have invested much effort in drug-treatment programs until recently, even though an overwhelming majority of inmates have some record of substance abuse.

Mr. Maxwell's original career plans did not include a fourteen-minute commute from his tidy home overlooking the Atlantic Ocean down a country lane and along U.S. 1 to a major prison. He and his wife, Diane, used to live in Boston, where their view involved the walls of next-door buildings and where he taught radiation physics to dental students. "I had the money, the prestige, the fast career track," he says. "Or I could come to a small town with a small community church and a slower pace. Guess what I chose."

Twenty-four-Hour Service Except for 11:30 P.M. Friday Until 5:00 A.M. Saturday and 11:30 P.M. Saturday Until 7:00 A.M. Sunday (Also Christmas Eve and Christmas)

ACCORDING TO ONE GRAVESTONE inscription in the old cemetery at Waldoboro, "This town was settled in 1738 by Germans who immigrated to this place with the promise and expectation of finding a prosperous city, instead of which they found nothing but wilderness."

Today, Waldoboro's approximately 1,100 residents (not to mention the thousands more who pass through on the highway) will find Moody's Diner, one of the nation's few surviving establishments that advertises home-cooked meals and means it. And if today's population of

Waldoboro (Samuel Waldo was another one of those Colonial generals who ended up owning most of a county) is barely half what it was sixty-odd years ago when Moody's first opened, just about everyone in town has eaten there and downed a tasty piece of pie they shouldn't have.

Thousands of restaurants have erupted across America; in the last decade, fast-food stores were the fastest-growing, up 78 percent, to 120,000 now. McDonald's alone has nearly 12,000! These "hurry up and eat because, well, just because" places line virtually every possible corner. On U.S. 1, they don't even wait for corners; anywhere will do—McDonald's, Burger King, Roy Rogers, Kentucky Fried Chicken, Pizza Hut, Dairy Queen, Denny's, Domino's, Red Lobster, one or two Armando's, and a fair number of Casa de la Somethings.

But unlike these chain establishments, Moody's serves mashed potatoes, mashed by hand from real potatoes. Moody's serves home-baked pastries with the same recipes that Bertha (Alvah's Mrs.) began perfecting shortly after they opened their first three tourist cabins back in 1927 before computer confirmation numbers were necessary. Each cabin had a screened porch (travelers stopped before dark in those days and sat down to relax and chat) and an outhouse out back. The price was one dollar per person a night, not including sales tax, because there was no such thing then. The price, however, did include a cold jug of springwater. For the first few years, heat was absent. Then each cabin got a wood stove (and the Moody boys provided room service for firewood each night).

With no restaurant, the Moodys sent their customers downtown to Brown's, until Alvah added gas pumps and a breakfast room and then a lunch wagon next door peddling hot dogs and hamburgers. When the new U.S. 1 opened in August 1934, on the other side of the hill, Moody's moved their lunch wagon back there. And that's where the diner grew, too, along with its reputation for home-cooked goodness and nearly twenty-four-hour service. Back before Interstate 95 sapped a good deal of the local lifeblood, truckers stopped there all the time (proving that theory for once), but so did local workers and businessmen and lots of families traveling to and from Bah Hahbor for vacation.

Alvah (his friends called him P.B.) had some funny notions about running a restaurant. He thought if you ran a clean place (he scrubbed the floors himself to make sure) with friendly help and good food at reasonable prices that was served fairly fast even without a drive-thru, then customers would return. They would come every morning, perhaps, for the coffee and gossip or every Thursday for the New England Boiled Dinner, or every Saturday for the beef smothered in gravy not from a jar, or on Sundays for the turkey, roasted upside down so the natural juices soak into the breast. He thought they would come for the fifteen kinds of homemade pies, including walnut, custard, fresh blueberry, and peanut-butter cream, which the Mrs. made starting at 4:00 A.M. daily.

And a funny thing happened: They did.

And they still do, some second-generation customers. Although some of the regular locals understandably shy away at peak tourist times like summer Friday nights and Sunday afternoons, Moody's has provided a comfortable living for 2.5 generations of Moodys. And since Alvah didn't believe in borrowing to spend, none of his nine kids did, either. So the place has never been saddled with debts. It still isn't, even with all the tax breaks designed to encourage it. Whether the children liked it or not, they also believed in hard work, each one of them putting in long hours waiting on tables or tending the stoves.

In return, they ate well, learned the business and some responsibility, and met some interesting people, not a few of them famous. Caroline Kennedy ate there once and agreed to be photographed with Eda Hoak, her excited waitress, who forgot to smile when the shutter snapped. One of those actors from "Gilligan's Island" stopped by. So did Gene Rayburn, a TV game-show host who was famous once. Sheila, one of the other waitresses, thought he looked familiar. So she walked right up and asked, "Are you the real thing or do you just look like him?"

"I'm the real thing," he replied.

But everybody gets the same food with the same service in the same comforting surroundings. So attached did the regulars become to Moody's that they vetoed the Moody family's plans to remodel the place; so the family kept the same awkward arrangement with the door by the cash register and settled for new tabletops and double-pane windows made to look old-fashioned.

Change has swirled around Moody's and Waldoboro. There's an electronics plant in town now making unrecognizable little things that go inside something else, which some of the workers making them may understand. The old pearl-button factory that used local shells now makes plastic ones. And even with the town sewer system complete, not many people dare to eat the clams they catch from the Medamak River; something invisible may be drifting down the river, because there are strange growths on the shellfish sometimes.

Moody's has managed to keep the price of fish and chips at $4.75, the fish rolls at $2.10, and a burger, fries, and a vegetable at $2.35. But Alvah's kids are middle-aged now. They still cook and tend tables and sell the family cookbook, *What's Cooking at Moody's Diner,* written by Nancy Moody Genthner, who still lives just down the highway. And they still don't take credit cards or checks.

But it's tough keeping Alvah's grandchildren interested in the business. "They're too smart to work on weekends," says Nancy, who's looking around for a professional manager to train in the Moody ways. "We're trying to hold on," she says, "still."

There is some resistance to change on 1. Wiscasset (in Indian, "meeting of three rivers") is an aging port with numerous graceful old homes that began housing a notable community of artists early in the century. Local legend has it that Samuel Clough, captain of the *Sally,* and an eighteenth-century merchantman engaged in trade with France, was involved with a plot to rescue Marie Antoinette from the perils of revolutionary France. His Wiscasset home was readied for the royal arrival. But the Queen of France didn't make it out of there. And one of the king's satin robes became a dress for Mrs. Clough.

Then, Wiscasset's elected council voted the other day to veto a plan to move two historic schooners to a maritime museum. Side by side at their riverfront moorings, the decaying wooden vessels—the *Hesper* and *Luther Little*—have been lifted off the tidal mud and gently set back down twice a day for the last sixty years. They're not going anywhere ever again and that's just fine with the folks who see them every day sitting there like a living memory. An entrepreneur had proposed moving the vessels to make way for a new marina.

During the American Revolutionary period, British authorities aroused considerable local animosity by marking the sturdiest pines—those with a minimum diameter of twenty-four inches—with a broad arrow mark, meaning they were reserved for the exclusive use of the Royal Navy as masts. Although shipbuilding has faded at many coastal Maine communities, it remains a going business at nearby Bath. But now, of course, the animosities are aroused by the modern vagaries of government budget cutting and *not* building ships. Over nearly two centuries, Bath shipbuilders, most notably the Bath Iron Works, have produced more than four thousand vessels and launched them in the Kennebec River. These craft have included everything from private yachts and lighthouse tenders to modern frigates. Bath did not build the famous old *Maine,* parts of which have rested since 1898 in the mud of Havana harbor. But Bath did build a massive ship named for another state on U.S. 1, the battleship *Georgia.* Launchings remain a major event just below the U.S. 1 bridge over the Kennebec, where, at night, the lights of the shipyard with its gangly cranes and the sparking torches of its welders provide quite a spectacle. Downtown, the old-fashioned store signs are painted, not neon, and the sidewalk clock, a vestige of eras when everyone did not wear their own watch, keeps accurate time.

Just beyond Bath, however, more than 450 miles out of Fort Kent, U.S. 1 goes modern, suddenly expanding into a four-lane thoroughfare past numerous commercial strips that are but harbingers of what is to come farther south—the Maineline Motel, the Siesta Motel, Texacos, Mobils, pizza joints, the Hong Kong Island restaurant, Satellite Auto Glass, a Country Inn, which is no longer in the country, and Bill Dodge's Oldsmobile and Used Car Annex. For a hectic two hundred yards, old 1 merges with I-95 before ducking off the high-speed highway into a country back road where the houses are far apart and each is lit by a lonely yard light standing in the cold.

Nearby is Brunswick, which due to its cheap water power once prospered as a mill town (cheap labor was imported from French-speaking Quebec) and as a sawmill town (once, there were twenty-five in operation). Their log disassembly lines helped produce the immense Tontine Hotel, a thirty-room downtown hostelry that cost fully seven thousand dollars to build in 1828 as the local rest stop for the through-stage along the coast road.

The community's original name was Pejepscot, which some wise city planners renamed for the elector of Brunswick, who became Britain's King George I in 1714. Modern Brunswick lost some historic old structures when the U.S. 1 bypass went through years ago. But it remains home to Bowdoin College, which was founded in 1794 and continued the time-honored American tradition of colleges naming themselves and buildings for someone who put up a lot of money (James Bowdoin of Boston). Bowdoin's class of '25 (1825!) produced such illustrious graduates as Henry Wadsworth Longfellow and Nathaniel Hawthorne, who is still dead, as evidenced by his writings. Not far off U.S. 1 at 63 Federal Street is the house on whose kitchen table Harriet Beecher Stowe, wife of a Bowdoin prof, produced *Uncle Tom's Cabin,* a novel whose powerful images helped unleash the fatal furies of civil war and turn plowshares into swords and many thousands of ordinary people into bodies. When President Lincoln met the author in Washington not far from some of the bloodiest battlegrounds, he said, "So this is the little lady who made this great war."

Last Chance for Peacefulness

FREEPORT (IN INDIAN, THE name does not mean "place by the water where the Chamber of Commerce meets for lunch on Thursdays to attract more busloads of tourists seeking bargains") is not a free port.

The community is a former town infrequently referred to as "the birthplace of Maine"; this is because the final papers for the separation of Maine from Massachusetts were signed at Jameson's Tavern in Freeport in 1820. Two decades later, they got around to drawing the northern boundary between Maine and Canada with the usual attention to local details of distant diplomats. As one result, there was no war with Canada. As another, not a few houses sit like Cecille Bechard's smack-dab on the border, with her refrigerator and back porch in the States and her bedroom and the northern half of her kitchen table residing in Canada. (When, according to an apochryphal story, Canadian officials informed her one day that the border had been adjusted a few feet and her entire house now sat within their jurisdiction, she moaned and wondered, But how am I going to handle those Canadian winters?)

The charter for the Freeport area's settlement was granted in 1680. However, owing to Indian hostilities and the penchant of these neighbors for kidnapping females, Freeport did not exactly experience a population explosion until after 1720. Slowly, however, settlers made their way up the King's Highway, as parts of U.S. 1 were once known because no king ever set foot on them. By the late 1800s, "summer people" began appearing like locusts. The electric street trolley from Portland came through in 1902, and to create a destination for more riders of his electric railroad, Amos Gerald built the Casco Castle, a huge hotel and amusement park. That's gone now, except for the chimney, but this is a strategy that has been refined all the way down highway 1 by the founders of Florida, who weren't from there, either, and more recently has been perfected by theme-park builders along many roads.

Freeport's real future, however, was launched with the simple mailing of one parcel of hunting boots in 1912. The mailee is unknown. The mailer was a businessman named L. L. Bean, who would do for boots what Richard W. Sears and Alvah C. Roebuck would do for watches, turn them into mass mail-order items and change the face of American merchandising, not to mention downtown Freeport and some other communities. In the process, these merchandisers and their 800 numbers also changed the wardrobes of millions of warm and dry American suburbanites with credit cards, who wouldn't know a duck blind from a .12-gauge over-under but, now that "country" was in, still wanted to look rugged, you know, but not raw, outdoorsy, but not rubeish.

Successfully selling such things from a small Maine town in a crowded market is no small accomplishment. But if some Americans, who never got closer to a mountain than a pile of potatoes at Thanksgiving, could be convinced to buy mountain-climbing gear or a different kind of canvas shoe for every sport, even something called "street-hiking," then there was hope for the future of Freeport's economy.

With suitable symbolism, Freeport's modern success actually rose from the ashes of a fire in 1981 at Edgar Leighton's Five and Dime downtown. Edgar had founded the store. Business had been good for his family in a time when a nickel or a dime could buy more than a piece of bubble gum. His store fronted on Freeport's busy Main Street, which was also the nation's busy Main Street. A fire like Leighton's would tie up traffic for hours on U.S. 1, which was the only way from here to there. For generations everyone—housewives bound for the grocery, vacationers bound for their distant cabin, honeymooners seeking solitude, trucks from Canada, farmers from up Waldoboro way, and the night shift at the shoe factory— had been funneled through downtown on old 1.

This was before even highways had to specialize. The same two-lane road would haul people from place to place at slow speeds or slightly faster ones, depending on what other vehicle or farm implement the fates plopped in front of you. Then, when the highway came to a town like Freeport, it would slow down and become a local street for a mile or two. And the passengers in the cars passing through would slow down, too, and look out the window at the community, which was fairly tidy because for reasons that no one ever explained but everyone understood, most communities wanted to make a good impression—even on people they didn't know and would never see again.

In summer, a banner would hang across Main Street announcing the annual local festival; every town needs one as a reflection of its personality. (And the Monday after the local holiday, the festival sign would come down for another year because the decorations committee was always among the most efficient in those days.) Travelers might eat at one of the festival booths or at one of the tables set aside for nonregulars at the local restaurants recommended by Duncan Hines in his annual guides to local dining.

If travelers needed help, they could stop for directions at a place called a service station, because the men there had grown up in the area and could speak an entire sentence in English. Service-station operators vaguely thought that by being friendly and helpful, passersby might turn into customers at their station or perhaps another one belonging to the same company down the road, where those operators were being friendly and helpful, too. Goodwill was just common sense then and usually contagious; it had not yet been refined to a strategy. But it did have potential.

If travelers saw Leighton's Five and Dime and remembered something they needed, they could just pull over and run in to find it on the wooden tables. And Edgar himself might even wait on them or wave from the next aisle over.

By more modern times, of course, simply pulling over was not possible. There were parking meters to time people and make sure that no one wasted too many chunks of twelve minutes lollygagging around. Every pulling-over place was full. Traffic moved very slowly because there now were stoplights to organize every corner. There was so much traffic and so little space for it on the old streets, and so many people driving up and down, the downtown blocks were just moving around until someone pulled out and a pulling-over spot opened up.

When horses controlled the traffic flow, they were smart enough to avoid collisions. Humans in autos, however, required experts and colored lights (red for stop, green for go, yellow for go *very* fast). Traffic planners decided that the downtown congestion could be eliminated by rerouting the through-traffic around the edge of town, leaving the downtown free for local traffic. What happened in many cases was that by rerouting so many people around the edge of town, traffic planners eliminated the downtown, leaving that area free of traffic congestion because it was dying.

This prompted a few officials to raise the price of parking meters, a logic that grows in the minds of some civic bureaucrats like refrigerator mold. Some towns declared meter-free days or removed them altogether, which did little for the business of parking-ticket writing. Still others brought in beautification experts who closed off a few streets and began driving buses around in case there were any New Yorkers moving to town, meaning large numbers of lost souls without a driver's license and proclaiming pride in it.

One nifty idea that was popular at the sparsely attended evening council meetings involved placing spindly trees in cement vats at strategic spots to create shade and the color green that Americans prefer on their money and so many other things except packages of meat. This plan didn't work, either, because in small towns no one in their right mind who hasn't just encountered an old friend stands around on a sunny sidewalk doing nothing but appreciating a little tree. Such a scene reeks of idleness, which long ago was banned by elderly aunts and ministers. And over the millennia, trees had become accustomed to living in earth, which could drain itself after a rainstorm and insulate the roots come winter. And even the most stylish prefab cement vat couldn't do that. So the downtown trees died, too.

On the other hand, the bypasses' roadsides became other downtowns with their more abundant, therefore cheaper, space for stores and vast tree-free parking areas. They were elongated downtowns and, unless town officials were quick at their somnolent civic zoning sessions, these new areas were actually quite ugly. Such development was great for real estate agents and enabled a few former farmers to move to Florida. The town governments, which received a share of the state's sales-tax revenues on every sale, not to mention the mushrooming property taxes on the valuable commercial land, were pleased because, among other things, the new income enabled them to employ more planners.

These commercial strips and sprawling malls on the bypasses became so popular that the traffic planners' sons had to install stoplights at every cross street to organize the traffic flow and speed it along. This worked just as well as the plan to ease downtown congestion. They added new lights with very short segments for left-turners and very long segments for pedestrians, in case any humans wanted to cross the street on their hands and knees. Eventually, these experts got each corner's complete traffic-light cycle up near the five-minute mark. They were also successful in ignoring the suggestion by one obviously deranged citizen to coordinate the lights corner after corner so that a vehicle might move, say, three blocks without stopping.

This necessitated invention of the urban expressway, which involved removing some old neighborhoods here and there. But a lot of those people didn't vote, anyway. Expressways would have no stoplights at all. Therefore, drivers could zip right along free of congestion.

Edgar Leighton remembers when the bypass went in around Freeport back in the early fifties. He remembers working at one of the shoe factories, which have gone out of business now. So when the fire hit, Leighton's moved, too, even though the town had put in brick side-walks and trees downtown. Edgar sold his place for a good price.

Meanwhile, L. L. Bean was changing its marketing strategy. It was happy with hunters. Trouble is, there aren't really all that many people willing to walk around in rubber pants and camouflage coats. So Bean went after the nonoutdoors outdoorsmen. The auto makers were successfully mining this suburban market with four-wheel-drive vehicles that say so on the side but are never actually put into four-wheel drive because then they would get muddy and possibly scratched. (There was one unconfirmed report that a woman in suburban Boston did put her family's Jeep Cherokee into four-wheel drive after a one-inch snowfall last February. "Thank God I had the four-wheel drive," she said that day at a baby shower. And her friends nodded.)

Bean's broadened its merchandise to kidswear and wifewear and daddywear. Bean's sold fold-up canvas stools to portage on those weekend safaris to the daughter's soccer game in the park, and fashionable jackets, slacks, and shirts in new materials with special washing instructions and funny colors like bayberry, which no normal hunter would be accidentally shot dead in. As one result, business boomed. As another, Bean's sizable parking lot, which had once been home solely to pickup trucks shouldering dirty fiberglass caps on the back, became the commercial refuge as well for Volvos, Audis, and Dodge minivans with fake wood trim and teddy bears snoozing on the backseat.

Freeport's attractions have long included the Mast Landing Bird Sanctuary, the ocean, some seals, and a visiting whale or two. But the Freeport Merchants Association began seeing factory-outlet stores with their apparent bargains as a potential bonanza. Bean's new downtown store, where U.S. 1 turns right toward Portland, stays open twenty-four hours a day 365 days a year for those nice customers who might want to drive all the way to Freeport and pay Maine's sales tax instead of ordering by mail or phone, tax free.

As one result, when the deputy sheriff makes the regular rounds about 1:00 A.M. even on a mid-winter Sunday morning, she thinks nothing of seeing diesel tour buses idling in their special curbside zones awaiting the return of their insomniac shoppers. Every conceivable parking space is filled with cars that bear out-of-state plates and college stickers and disgorge people to be drawn like moths toward the brightly glowing windows of Bean's nearby. Peering in, one can see not burglars but dozens of snug strangers strolling all the floors, examining sweaters and kayaks, flannel pj's and duck calls, and beautiful suede boots that are absolutely, positively stain-resistant, unless you're going to wear them outdoors.

Once upon a time, many towns dreamed of having a major mall as income generator and job provider. It would restore the community's identity as a destination in itself, a kind of self-enclosed, climate-controlled downtown (even to having indoor vats of trees, which are watered after closing by maintenance people). Malls also provide the kind of prideful commercial recognition of a community's existence that post offices or army bases once did. We have a mall; therefore, others see that we exist. Many communities must make do with a Dairy Queen, which isn't exactly national TV. But they *can* claim to have exactly the same ice cream as any of the more famous places. (And if the community's people knew that the larger malls can become such magnets of mischief that they are set aside as self-contained police precincts, then the normal residents might be real happy to stick with their mail-order lives.) From seeing a blur of these communities over many years of travel, it often seemed to me that they are wonderful places to live, but you wouldn't want to visit there.

On the other hand, Freeport, which in some ways has turned the entire town into a mall, may be a precursor of a new kind of tourist destination: one that many want to visit while it remains a lovely place to live, so long as you're not trying to find a parking space anytime during the month of August. That is the busiest month for acquiring in Freeport, although the Christmas season is very good, too. (Christmas in mail-order land begins when the December catalogs go out right around Labor Day when the Back-to-School sales are over.)

So successful has Freeport's strategy become that 85 of the 102 members of the Merchants Association are factory-outlet stores. Leighton's became a Dansk store. London Fog came in. So did Calvin Klein and Van Heusen and Ralph Lauren and Reebok. Timberland boots is there, too, although its image is more associated with Minnesota. And Alpine sheets and Hathaway shirts are there, as well as every conceivable kind of women's shoe designer. (Cole-Haan took over the old Western Auto store.) By the beginning of the 1990s, every downtown lot, save two old houses destined for destruction, contained a retail outlet. The Board of Appeals even allowed McDonald's to open in the old Gore House as long as it complied with storefront restrictions against those big ugly arches. Any further outlets are steered south of town out on U.S. 1.

Freeport's annual retail sales now total in excess of $180 million—not too shabby for a discount town with only seven thousand people who stay longer than overnight. Before the freeze prompts Louis Marstaller to turn off his plumbing, a few of the visitors stay out at his place, the Maine Idyll Motor Court. Louis's folks started it back in 1932 when the phone number was 90 and the boys rode each customer's running board to show them their cabin back in the trees. Today, if all the beds in all twenty cabins are full, the old place can take sixty guests, not counting their pets, which more people travel with and which the motor court allows.

Rates at first were the same as Moody's—one dollar per person. Now the cabins cost from thirty-eight dollars per night all the way up to seventy-five dollars for three bedrooms and a kitchen—plus Maine sales tax, of course, which is 5 percent on merchandise but jumps to 7 on motel rooms since, legislators figured, those people are

from out-of-state, anyway. And since even dinky little Connecticut down the old highway charges 8 percent, Maine still can seem like a bargain. But then Easterners— Northeasterners, anyway—don't seem to get out and compare things as much as other regions' residents. Northeasterners figure it's their birthright to pay more than necessary for an aging infrastructure that, like New York City's, seems to spring major leaks just about every Monday-morning rush hour. It is hard to imagine, for instance, Rhode Island residents mounting a tax revolt like California's (or anyone hearing about it, if they did). Or Massachusetts legislative study groups pondering how Montana, out west on U.S. 2, has made it almost to the twenty-first century without any sales tax.

Although the Maine Idyll appreciates your business very much, it does not take credit cards. Louis will, however, accept a personal check. "We're not in the big city with lots of bad things going on," he explains. Guests seem surprised at the check policy, but Louis isn't. He gets about one bad one a year, which, as it happens, is still cheaper than paying 3 percent to a bank card at some postal box in Delaware. "We don't make a lot," Louis says, "but we get by."

He also runs into some pretty interesting people, including parents who stayed at the Maine Idyll with their own parents many years ago. For such repeaters, Louis gets out the old register book and finds their father's signature back on one of the 1939 pages or somewhere. "It's fun," he said.

And, since the outlet business took off, the business of providing beds to exhausted spenders between shopping sprees is pretty steady, too. Last year, more than 3 million people made the pilgrimage to this commercial mecca. That's about a quarter million more people than visited Yosemite National Park. In a couple of years, Freeport's merchants can expect their stores to surpass the old Grand Canyon as a grand, modern American tourist site.

Another Good Place to Live

IT WAS GOOD OF the early white settlers to choose a site for Portland that placed that major market so close to Freeport's stores. Some history books mistakenly attribute the location of Portland (in Indian, Portland is not a word) to the existence of Casco Bay (in Indian, it means "place of the herons"). The deep, peaceful bay was also a popular gathering place for explorers who wanted to reach the North American continent without actually having to mingle with the natives.

At first, the Indians were friendly toward the newcomers, even without a chamber of commerce. Although the foreigners did talk funny and wear lace, both groups favored feathers for headdressings. The Indians were not into powder and perfume quite so much, but then they didn't have to live with each other below deck for two months to reach Portland. The newcomers, at least initially, were not too good in the woods, their red and blue coats failing to fool many wildlife. The newcomers seemed to go for the caribou just fine. However, they had the habit of not paying their bills with any good goods or offering less for the lifesaving stores than the Indians thought was proper. Over the centuries, the Indians had developed a certain proprietary feeling about the land, and were of the opinion that they were the settlers and whites need not apply. This attitude did not jibe with London companies issuing the same land through grants to close friends; sometimes, in the spirit of greed, awarding the same land twice. There was, too, a fair amount of woman-taking by the Europeans, who would eventually write the history books and forget to mention this part of taming the frontier. There was, too, a fair portion of lying and cheating and insulting and not a few shootings of Indians who had many questions about this process of civilizing a wild land.

Strange as it seemed then, such incidents alienated many of the remaining Indians, such that after the first few boatloads of explorers were welcomed, the others deemed it better to anchor several arrow flights' distance offshore. Casco Bay also was equipped with numerous islands, which were scenic and, perhaps more to the point, uninhabited. These islands eventually became popular destinations for steamboats full of tourists, and their beaches provided perfect fodder for tales of pirates and other frightfully interesting characters (CAPTAIN KIDD BURIED TREASURE HERE, TOO). Jewell Island (George Jewell, an early white owner) was the focus of treasure seekers using divining rods, good-luck charms, animal sacrifices, and known idiots believed to possess special powers. Cliff Island was inhabited by men seeking to put treasure on the beach; they lured ships onto the rocks and then collected the salvage. Harriet Beecher Stowe, who helped found the Civil War, resided near the ferry landing on Orrs Island for a time and made it the scene of "The Pearl of Orr's Island." Later, Admiral Robert E. Peary lived on stony Eagle Island, when he was trying to get away from a busy time at one of the Poles.

White people began living in Portland around 1630, the same year John Winthrop's flock arrived on the *Arabella,* and the usual lumbering, fishing, and shipping businesses eventually erupted. However, development was spotty at first, owing to frequent attacks by Indians, by French-backed Indians, by French-backed French, and eventually by the British themselves attacking during the Revolution. Much of the city burned when a July Fourth firecracker set off a major conflagration in 1866, a date that explains the Victorian flavor of much of the reconstruction. The city has been home to Henry Wadsworth Longfellow (before Bowdoin) and Neal Dow, who was a

Civil War general better known as the Father of Prohibition. Today, Portland, which has a median age of a surprisingly low thirty-three, has become a popular residence for migrating men and women with portable job skills like medicine and law, and who favor safer city lives and the smog-free outdoors, which, in Maine, are never more than a few blocks distant.

Nearly five hundred miles out of Fort Kent, U.S. 1 runs through the downtown past the Frederick Payne Federal Building, Havelock Field, the Exposition Center. It turns left at the railroad tracks to become St. John's Street, paralleling I-95 and passing McDonald's, Dick Cole's Tire Center, Lisa's Pizzas, Burger King, Colonial TV, Countryside Butchers, the Greyhound Bus Station, Handyman Equipment Rental, and the Maine Memorial Company, which serves the nearby cemetery with engraved stones.

South of downtown as property grows cheaper, a few small old homes still sit, punctuated by Rudy's Lunch Counter, more pizza places, a Day's Inn, and some small strip malls adjoining the Portla   rive-In Theatre, one of many old outdoor theaters whose sign has seen better days—and more customers. Next door is another sign of the United States's changing social landscape, the Tom Thumb Daycare Center, named for a nineteenth-century sideshow freak who has now been cute-ified in the American mind (and who lived on this same highway). Owned and licensed by a beauty-school operator, Tom Thumb is open eleven hours a day from 6:30 A.M. to handle the infants and youngsters of single and dual working parents. They pay from $105 per week for the wee folk age twelve months to $90 for those at least three and a half.

The school, one of many such private enterprises that have sprung up to replace the mothers and grandmothers who once stayed home for such chores but now can't or won't, is licensed to care for fifty-six youngsters in the former lumberyard building, according to Vida Wronski, the director.

Nationally, the Bureau of Labor Statistics reports that the percentage of working-age women now employed or actively seeking work is hovering at 58 percent. That's up sharply from the 37 percent of 1960, though still below the 75 percent of working men, a rate that is gradually declining. Ms. Wronski and Tom Thumb's five teachers use a more homespun predictor of future enrollment: They watch the price of home heating oil. When it goes up, more Portland spouses seek work and the daytime population of the Tom Thumb Daycare Center rises, too.

Ms. Wronski stayed home until her child was ten. She felt he was more intellectually involved as a result but had a harder time at first mixing with other children. "I think when parents stay home with kids," she said, "the adults seem more eager to get away more often." This situation is a potential buried treasure that many hotels are trying to mine on their slower weekends by offering special "getaway packages."

To ease any weekday adjustments by the youngsters—and their anxious parents—the center has an "open-lunch" policy. If desired, parents are allowed to visit their child for noontime sandwiches. "We want to make this as much like home as possible," Ms. Wronski said.

The center is also considering offering evening care until midnight for the growing number of families who have someone working that late shift. One way or another, they all drive on Route 1 past Jolly John's Used Cars, the Sunrise Motel (WEEKLY VACANCY LOW RATES COFFEE PHONES), and some of its country-cousin cabins and motor courts from the 1930s. If their eyes are not glazed, they also encounter the York County Pistol Range and Gun Shop, the Vacationland Bowling Center (COME ROCK AND BOWL), Captain Isaiah's Haunted Mansion, and, some two thousand miles from the Caribbean, Burrito Junction and the start of Spanish foods. Nowadays, the gas stations have become general stores again, the barbershops are unisex hair salons, and the elderly gentleman helping elderly women across U.S. 1 for early mass at St. Joseph's in Biddeford looks as if he could use a little help himself.

Biddeford on the Saco River was the sad site of an

encounter in 1675 between some drunken sailors and an Indian woman who, with her baby, was rowing in the river. The men desired to test a theory in white society that Indians could swim by instinct from birth. They tipped the canoe. The child disproved the theory. And for years afterward, his father, Chief Squando, wreaked havoc and revenge on settlers. Once things had settled down, much of Biddeford was peopled by what one old guidebook calls "respectable Scottish immigrants." So there were some. Today, Biddeford is home to one of the state's four incinerators, whose flames consume nearly half the state's annual production of 1.4 million tons of solid waste.

Traffic around the quaint, not-so-little-anymore Kennebunks was always bad in summer; you wouldn't believe the odor of all those horses and their by-products, which is why stiff sea breezes had to be invented. But it (the traffic) got even worse after George Bush was elected president in 1988. Mr. Bush is either from Connecticut or Maine on U.S. 1 or from Texas, which wouldn't be caught dead with such a slow lane. A fair number of hardy retired couples like Henry and Marjory Spitzhoff, who like sand but don't need ninety-eight-degree days with matching humidity, have settled here after busy lives elsewhere. Many visitors, famous and not-so, have enjoyed the shady elms and nautical airs of Kennebunk, Kennebunkport, and Kennebunk Beach, including Kenneth Roberts, who was born there and then wrote *Northwest Passage,* and Lafayette, who was not born in Kennebunk but stopped by anyway (LAFAYETTE ATTENDED A RECEPTION HERE). So tied to the ocean are all these Kennebunks that Santa Claus makes his arrival on a lobster boat to launch the Christmas shopping season.

President Bush was not the nation's first chief executive to use Route 1; the first was the first. But Mr. Bush's summer-home compound in Kennebunkport is not far off the highway, down by the ocean, where the gray waves from Europe crash up against the gray rocks of North America and, now, trim men wearing dark sunglasses, well-cut business suits, unbuttoned, and flesh-colored earphones aim their stares steadily out to sea—and occasionally mutter something into a shirt cuff. Likewise, the president's off-season white home is just a couple of desolate blocks off the same U.S. 1 down in Washington, the former swamp that, no doubt, just happened to be near the first president's estate at Mount Vernon, a bit farther down the road.

Not too many visitors get to see a president, wherever his vacation villa happens to be, although a large number claim to have heard his helicopter. So the president's partial presence nearby has not done a whole lot for Jim Brewster several miles down the 340-year-old King's Highway in Ogunquit. The area was purchased in the mid-1600s by a gaggle of liberal religious refugees from the Cotton Mather school of intolerance then reigning over Massachusetts. King's Highway was often used by religious refugees in those days since it led both ways out of Boston, up toward Indian country and down toward Rhode Island, which had its own set of rules, but burning people was less popular there. Fleeing seems to have been a very popular pastime back then. One of these notorious heretics passing through Ogunquit was named Anne Hutchinson, who would later provide the name for a popular parkway that crosses U.S. 1 on the New York City line—and charges every car twenty-five cents.

The town of Wells, not far from Ogunquit, once was indicted for not having a school. But Ogunquit, without even knowing that the name means "beautiful place by the sea," evolved into a popular seaside resort and home for a recognized community of artists. The area's beaches are long and lovely. And the highway is long and jam-packed. John Brown, a retired mason in Wells, can remember when only one or two automobiles a day interrupted the steady flow of horses, who are now seen only in parades. The riding stables went out of business because people, even those allegedly on vacation, were in more of a hurry and didn't feel they had the time to sit atop a large animal strolling along at its own pace. So now they bring their own horsepower and sit inside them for long traffic jams on a crowded coastal highway that is filled with others sitting in their vehicles feeling hurried. One traffic planner was quoted in the *York County Coast Star* as blaming U.S. 1 for being unable to decide whether it wanted to be a through street or a local road.

Jim Brewster's little unheated office sits off U.S. 1, *just* off it. The shack was way off the highway until 1932 when FDR's WPA hired unemployed men to widen the young road. Ever since then, Jim's chair sits just ten feet from all the traffic, which is fine with him. Jim runs Brewster's Taxi. So did his father, and his father's father, and his father's father's father. Originally—in America, anyway—the Brewsters were from Plymouth Colony. Actually, they came through Plymouth. They didn't stay there long. As soon as they stepped off the rock, there were troubles with the authorities, something about working on Sunday. This didn't seem to be a problem in Maine; in fact, the Brewsters are so well thought of in Maine that the Ogunquit volunteer fire department has named its old truck after Jim's great-great-granddaddy. By the late 1800s, the Brewster name was synonomous with taxi in Ogunquit.

The Brewsters have hauled thousands of people and tons of luggage over the years. Well, in the early years, it was the Brewster's horse who did the hauling. But by the early 1900s, the company had its first Ford. They still have one, but now it's called a Taurus and it's the spare that April or Gloria, Jim's daughter and his wife, drives on really busy days. Jim wheels around the big old Buick Electra station wagon.

Jim actually lives up the road in Portland. You can't really make a year-round living anymore driving a taxi in a community whose economy revolves around the beach. Not in Maine anyway. Maybe you can down in Florida, but there, the slow times are in the summer. Every spring, Jim, who teaches high school social studies on the side, starts checking the Ogunquit beach. When there are more people than sea gulls, he opens the taxi business for the season until the tourists travel south for the winter. "Without tourists," says Jim, "we'd have to eat dandelions and camp out up here."

Even in the summer, things are a little tougher since the Chamber of Commerce launched a fifty-cent shuttle bus around town. It tries to ease the traffic a little, which is especially horrendous when accidents happen, as they do so often with so many vehicles in such a confined space.

The customers and the business have changed, too. When Jim was little, he helped his grandfather muscle the ponderous steamer trunks off the baggage car of the reliable Pullman train making every stop from New York to Boston to Portland. The train passengers, who became Brewster passengers, were well-to-do city people coming to rent a cottage or entire house and mingle with the artists for a month or two at the seaside, where they sat in those low wooden-plank chairs with the arms that could double as a table for a glass of iced tea and a book. Life's pace was faster than the horse then, but no quicker than the trains. Dinners were planned for their culinary and conversational content. There weren't many portable stereo systems in those days.

Back then, it was an all-day taxi voyage to Boston and back on *the* road, which was U.S. 1. Today, with I-95's four lanes sometimes moving as slowly as the speed limit, it's a little over three hours, plus seventy-two dollars. It's about one dollar per minute for the forty-minute trek to the Portland airport, but that can be for up to six people. Airplanes seem to be the favored transportation for the well-to-do who still patronize Ogunquit. Although with all the airport parking and the traffic and the crowds and then the flight delays and the airport crowds and the traffic at the other end, a fair number of Ogunquit-bound New Yorkers have found that it is actually easier, usually faster, and certainly cheaper for a family to hire Brewster's Taxi. For the haul from, say, a Manhattan apartment to an Ogunquit beach motel, a vacationing family of up to six would need about six hours, including lunch, and $260. That's less money and time by cab than a couple of tickets on the airplane shuttle. Some of these people have been Brewster passengers for two or more generations, watching Jim grow up and then their children watching his children grow up.

Jim tries to keep his business small because he likes to know most of the customers. Inevitably, he engages them in conversations about all kinds of things. "We have a ball," he says. "I tell you, the variety of people you meet. The politics you hear. With the president just up the road. The things you learn. What people are thinking. Their minds are so interesting."

The clientele has changed, though. The ease of transportation, modern highways, affordable cars and lodging, social affluence, institutionalized vacations, and television, the great homogenizer, have combined to erase much of the exclusivity of many such vacation Valhallas. Pretty much everyone can come now, which means that dinners are less formal. With satellites and cable TV, what happened a few minutes ago in Washington is instant conversation fodder in Ogunquit. Today's tourists don't walk or ride around town anymore hoping to glimpse some of the famous folk who rented summer homes. (Now, famous folk favor condos in more distant climes.) And since most of Ogunquit's visitors have their own vehicles, few of them telephone for a taxi (since there's only one in Ogunquit, standing on the curb and trying to hail one could take the better part of a day, especially if Jim is on an airport run). Tourist couples take taxis locally now only when they're caught at the beach in a sudden downpour or want to return to their room with an armload of packages after an afternoon's exhausting expedition to the stores, which vacationers call shops. The men are still the ones who pay the taxi fares, and the tips seem to go in inverse proportion to the number of parcels the wife has him carrying.

Ogunquit's visitors tend to stay just a week or two now. "People are in more of a hurry," says Jim. Even when there's plenty of time, they're checking their watches and wondering out loud what's taking so long. Just seeing and hearing the helicopters *whup-whupping* their way to and from a president's compound seems to add an overlay of urgency to life, which is exciting for a while. Some of the vacationers' cars even have cellular telephones inside, their telltale little antennae on the trunk lid, where the rumble seat used to be, telling perceptive passersby how instantly in touch their drivers can be, which some people think is important and others smile at.

By early October, however, those kind of people are gone. And when, on some weekend drive by the beach, Jim notices the gulls outnumbering the people, he locks up his little shack by the side of U.S. 1 and takes the highway back north for the winter.

John Paul Jones Slept Here

MEANDERING SOUTH ON 1, the last incorporated town in Maine was once the state's first, Kittery, which confronts New Hampshire just across the Piscataqua River. A drawbridge, painted blue-gray and bright orange against insinuating rust, spans the river, linking Kittery with Portsmouth, New Hampshire's lone seaport. The first thing travelers see there sitting high and dry in a roadside park is the U.S.S. *Albacore,* No. 569, a Cold War submarine that never saw combat now converted to an attraction for tourists with their little cameras, sensible shoes, and simmering sibling feuds in the backseat. The next thing travelers see in New Hampshire is the Fifth Wheel, which features adults books and novelties.

Before the American Revolution, a trip from Portland to Boston could take three days by land. By 1753 when Benjamin Franklin was appointed postmaster general, former Indian paths were being organized into post roads to handle the Colonial mails; a few of Franklin's old stone mile markers have not been bulldozed or stolen. Franklin would be fired by the Crown ultimately for his independence, but to this day parts of what became U.S. 1 are better known as the Post Road. And there were not three lanes on December 13, 1774, when Paul Revere made good time on the highway to bring word to Portsmouth's Committee of Public Safety that the wary British were cutting off gunpowder shipments to the American Colonies. In what is reverently remembered, since it was successful, as another one of the first overt acts of the Revolution, the Sons of Liberty promptly attacked the local fort and captured five tons of gunpowder, much of which was spent at the Battle of Bunker Hill.

Today, on the Interstate, of course, the same Portland-Boston journey lasts little more than an hour. On U.S. 1, laden with local traffic, the trip can take somewhat longer. But then the interstate doesn't have fortune-tellers and Mel Clark's place and all the memories of a rich past.

In the early days of settlement, back before American citizens were so good at taking things like easy transportation for granted, communities like Portsmouth were largely self-contained and enjoyed at least a regional reputation based on some local resource. That reputation may have faded as the years passed, though the underlying economics often remain. The easy access to berries was a tasty, but not principal, bonus for founding in 1630 what would become one of America's oldest urban sites, even if Colonials had different ftandardf for their fpelling of Portsmouth's Strawbery Banke neighborhood. (And even if the Spanish were running around down in Florida settling St. Augustine along the same future highway.) A succeeding generation of residents, however, had the wisdom to fight when that influenza epidemic of urban renewal began spreading in the 1950s; they kept the narrow lanes and old homes of Strawbery Banke as a ten-acre outdoor heritage museum. And now that the bikers are gone, the streets are full of *patisseries* and cheese shops for riders of ten-speeds instead of Harleys.

Portsmouth's fame attracted all the usual Colonial celebrities, including James Madison and Lafayette. St. John's Church has even preserved the chair used at the November 1, 1789, services by the nation's first president (WASHINGTON SAT HERE). That initial fame of Portsmouth's was built on the city's access to wood for building ships, to water to put them on, and to the nautical and mercantile skills of its residents, whose prosperity financed so many elegant Georgian and early Colonial homes.

Portsmouth was, as historian Samuel Eliot Morrison put it, "the mast emporium of North America." The *Ranger* was built there. So was *Old Ironsides,* the *Constitution,* and *America,* the frustrating dream boat of Captain Jones, the Revolutionary War hero who never got to sail her. Today, nearly two centuries after its founding, the Portsmouth Navy Yard, which is really on Badgers Island off Kittery, still turns out ships for the nation's navy, though now it specializes in ships that intentionally go underwater. The yard was also the site for the signing of the 1905 treaty ending the Russo-Japanese War, which was not fought along U.S. 1.

On Middle Street near U.S. 1 stands the genteel boardinghouse run by the widow Purcell, where John Paul Jones resided for more than a year while overseeing construction of *America.* He arrived in Portsmouth, with his steward, on the last day of August 1781, after a nineteen-day journey by rented carriage up the Post Road from Philadelphia. According to his meticulous expense accounts, the carriage, three horses, and driver cost Jones eight dollars a day. The boardinghouse set the nation back ten dollars a week, but that included food, clean linen, and keep for Jones's steward, too. The ongoing unpleasantness with the British, who held New York City, caused the trio of travelers to detour slightly and meet in White Plains, New York, with General Washington, who had not yet given his name to a large bridge across the Hudson River at Fort Lee.

Jones had resided in Portsmouth before, in 1777, while outfitting the *Ranger.* That ship was named for Roger's Rangers, a band of New England guerrillas much revered for their wily woodland exploits, although Major Rogers saw fit to switch to the British side. The name *Ranger* would live on later in the form of an aircraft carrier launched in 1933.

But Jones's visit to Portsmouth in 1781 was to spur the languishing construction of *America*—at 182 feet, the largest ever built there. Jones berated the construction engineer, beseeched his friends in Philadelphia for more money, wrote other acquaintances to seek deckhand recruits, and helped track down a cheaper grade of canvas for sail. Dashing, deceitful young devil that he was, Jones also composed poetry, which he faithfully dispatched to every one of his true loves. Owing to the pressures of his work in Portsmouth, however, Jones had no time to write different poems for different ladies. So he merely changed a line here and there to match her nationality. Perhaps some of these verses were written by lantern light in the shipyard where the hardworking captain stood his turn at night-sentry duty when rumors had the British planning a raid to burn the new craft. Indeed, His Majesty's frigate *Albemarle* was lurking offshore, commanded by a twenty-three-year-old captain named Horatio Nelson, whose naval fame would come another day. Then, just as completion neared, the fate of *America* was sealed in Boston.

Despite dense fog, a French fleet returning from raids against the British in the Caribbean sadly followed the advice of a Boston harbor pilot anxious to get home. Three of the thirteen ships, including the immense *La Magnifique,* ran aground. The large ship broke up, with the loss of only a few cannon and considerable pride. But an American Congress, in a sign of gratitude and, perhaps, an early inkling of budget-cutting pressures on the military, gave away *America,* its sole ship of the line, to the French. (Portsmouthians still worry about the budget wars: the survival of Pease, the local Strategic Air Command base that could launch Armageddon.) Without notable complaint, the obedient Captain Jones presided over the tricky launching of *America* in 1782. Then he handed command over to the French, checked out of the widow Purcell's, and made his way back south on the highway. (Just as well, in a way, because within four years French

naval inspectors found the gift riddled with rot, having been built in too much of a hurry with unseasoned green timber. They ordered the craft broken up, which didn't do much for New Hampshire's shipbuilding reputation. To cover this up, some suspect, Portsmouth's shipbuilders instigated the French Revolution.)

The ships turned out in Portsmouth today (and the electric power generated at the nearby Seabrook and Wiscasset power plants) have more to do with nuclear fission than wood. So, not far away on U.S. 1, a band of hopeful state officials labor to teach modern Americans a bit about their one major renewable resource. The Urban Forestry Center is not a contradiction in terms. Presided over by J. B. Cullen, the state's administrator for forestry information and planning, the conference center is tucked into a 160-acre forest donated in 1976 by John Elwyn Stone, a descendant of New Hampshire's first governor (they keep track of such things in New England).

Using gardens and nature trails and indoor exhibits and lectures, the forestry center labors to teach urban Americans about those tall, woody things that grow silently by the millions just outside their consciousness. Mr. Cullen finds that most people know about maple syrup, of course, but few are aware of trees' role in processing carbon dioxide into oxygen, of trees' soothing influence on human spirits, and of trees' role in deadening the din of busy streets. "So much of our lives are always changing," says Mr. Cullen, "and change stresses people. But, you know, trees are the only things that change that people like. They even travel hundreds of miles to watch trees change."

At present, New Hampshire is copiously covered by trees, second only to Maine in tree cover. And nearly 86 percent of New Hampshire's millions of acres of trees are privately owned. "Many of the decisions affecting our heritage of trees are private," said Mr. Cullen. "We want those decisions to be right for the future. People don't come to New Hampshire to see highways and factories and clearings."

The center teaches teachers to use trees and leaves in classroom counting, not just apples. It trains loggers in safety and environmental consciousness. It provides lessons in woodlot management to owners of small plots and displays shrubbery exhibits for homeowners. It's not an easy job. There are more visitors every year asking questions, jogging on the tree trail, and even sitting for thirty minutes in their cars in the shady parking lot to munch their lunches where they can't hear any traffic. But vandalism is worse, too. "Signs, benches, everything gets destroyed," said Mr. Cullen. "If it's not anchored in four feet of cement, things have a way of walking out of here." Some people even slash the trees.

"Most Americans live in cities now," he said. "We want even those apartment dwellers to think about trees, because if we don't, someday trees won't be around much anymore and people will say, 'Why didn't we do something before it was too late?' As a people, you know, we don't really appreciate what we have. I had some European tourists in here last fall. They drove around the state. You should have seen them. They were amazed at the scale of our state's forests and the colorful splendor of all the different kinds of trees. They asked me if we had mixed them up like that on purpose when we were planting so they'd make a tourist attraction every fall."

One of the distinctive features along U.S. 1 that has nothing to do with trees or the Revolution (at least the American Revolution) is the sign at Mel Clark's place. The sign consists of a car, a 1951 Morris Minor, sitting maybe thirty feet up in the air on a post. Mel put it up there a quarter century ago (he took the motor out first) to advertise his used-car business. Although no one is known to have counted the countless double takes of drivers encountering a car considerably higher than the roadbed, it did seem to help his business. "How about that?" he says proudly. "It's a real eye-grabber, isn't it?" Mel is more into used trucks now. But he won't take the sign down until it rusts away.

Going for the eyeballs of the passerby is a hallowed tradition on the old road; it has worked for New Jersey FLEA MARKETS!!! and Florida ALLIGATOR FARMS!!! (LIVE!!!) And if thousands of city fathers (and mothers) have worked diligently to clean up the highway's roadside image in more recent decades, they have had to contend with brighter lights and grandfather clauses that protect some of the existing commercial, uh, sights. "I'm gonna put a big old Mack truck up there someday," says Mel. "Of course, with all the zoning laws, I'll probably have to do it at night. I'd like to put a truck trailer up there, too. Saw that down near Chattanooga one time. But maybe that's going too far. What do you think?"

Also available for display is a thirteen-and-a-half-foot-tall tire, which Mel knows is the world's biggest. "At least it was in the seventies."

Mel is a salesman, which becomes a tradition on many main roads as soon as the highwaymen get squeezed out and the second inn comes in. But Mel is not one of those salesmen who interrupt late-night TV movies to throw on a three-cornered hat for their superspectacular July Fourth sale of furniture that looks to be made of balsa wood. Mel sells motor vehicles. I can remember as a child in the late forties and fifties accompanying my parents on auto-hunting expeditions in showrooms where I could never figure out how they got those big cars inside. Those were the days when one company could name their fanciest vehicle the New Yorker and no one would laugh, because they did not picture all New York cars parked for more than four minutes up on blocks and minus their wheels. In those days, even in the big cities, your salesman (you belonged to one then as you did to your doctor and insurance agent) made a point of remembering your name from the last purchase. He would check in by phone now and then during the life of a new car to see how it was going, tell you what had happened to your last trade-in, make sure you remembered his name and interest in case you were considering owning two cars, which more people were doing now, he said. He might even pull in the driveway some Saturday in a shiny new convertible he just happened to be taking through the neighborhood.

For a while every fall, the auto showroom windows would be papered over to make you want to see what new-model chromed secrets were hidden within, awaiting the official day. If you knew the right people or if someday you might possibly consider buying a car, you might get in early to see the new wings or grilles or whatever on that year's cars, which could become an element of instant status in school-yard conversations, as long as everyone hadn't been allowed in. (Or you could simply watch the new cars being unloaded from the trucks the week before.) At night, the scheduled unveiling of the new cars was announced by the powerful beams of spotlights piercing the city's black sky as if seeking German bombers.

It was a very exciting time, so important that those childhood years are divided in my mind into eras according to what car we owned; there were the Plymouth years and the Dodge years interspersed with a brief Mercury span, and then, when they could see the end of the college bills, the Chrysler era. As a boy, though, if I was lucky enough to have a father who bought a new car, as I was every three years, then that first ride was like Christmas in October. The new car, which was partly mine because I rode in the backseat a lot, had a wonderful new smell (which the Japanese have patented now to spray on their new stereos just before shipping). The pristine cleanliness of the wide white tires. The look of the sleek dash. The sound of the new radio. The gliding sensation of the new ride. The endless possibilities of new life with a new car, like enjoying a newborn baby—only you can buy a new car on time.

Of course, to break in the new motor, we couldn't exceed thirty-five miles an hour for the first five hundred miles. That was very embarrassing, because we went so slowly that even the rusting clunkers with West Virginia plates passed us. But eventually, the break-in period ended, and Dad could step it up to fifty. It was his life's dream to buy a red convertible, which he did one year. With the top down, I would sit in the back in what quickly became a vicious wind tunnel. Nothing could be loose on the backseat or it was gone. The noisy blasts of air would buffet my ears and do nothing to my flattop, and when Mom turned around to issue some warning, all she

seemed to do was silently move her lips. It was just great. I dreamed that someday I would have my very own car: shiny, deep-throated, low in the rear, waxed by someone else. And I would be the one to set the radio buttons.

One time, Dad bought a used car. He didn't take Mom and me with him then. Used-car buying was like going into certain parts of town—not for children and bad bluffers. Used-car buying involves very hard bargaining, which is what Mel Clark likes about the business on U.S. 1 south of John Paul Jones's old rooming house. Mel remembers the old days, too. He's been selling used vehicles since recess in the seventh grade when he could shape nice deals on used bikes. He moved up to scooters, then cars, now trucks. "It was different then," he recalls. "You didn't need titles for anything, just a bill of sale. No sales tax. Now we've got rules about rules. I get inspectors in here for this and that. People running around with little meters checking you. If we don't have a war going on, then we end up watching each other all the time. And I've got so many people's hands in my pockets, there's no room for my own."

Five generations of Clarks have lived and worked along U.S. 1. In the thirties, his grandfather ran an auto-repair shop/gristmill/home-construction business just down the road, at a time when a valuable piece of property might bring five thousand dollars an acre. Now his grand-father's old lot is a discount TV store.

Mel remembers every summer when the gypsy junk peddlers in their horse-drawn wagons would come back from wherever gypsy junk peddlers went in the winter. All the youngsters' mothers would warn them to run and hide because gypsies liked to steal children. It was a known fact. Now the gypsy men seem to have settled down in the business of hot-topping asphalt driveways and the women tell fortunes, although just down the road in one of perhaps two dozen such establishments along U.S. 1, Duchess (PALM READINGS AND TAROT CARDS) does not grant interviews along with her peeks into the future; writers with notebooks can disturb the continuum, also business.

The tourist-cabin business has suffered over the years, too. Not so many Slumber Rest Manors, more Ponderosa steak houses and mattress cities, no more Studebaker dealers just Acura showrooms, even little warrens selling(!) the baseball cards kids once traded from their red wagons. The steam-train rides back into the woods down from Mel's shut down in the sixties. And commercial property along 1 can cost a half-million an acre, which is not necessarily what Mel paid Muriel Fogg for the land next to his when her husband died. Like a lot of farmers' children, the Fogg kids did not dream of sixteen-hour days farming, and moved away to a city— Mel thinks in Michigan somewhere. So the land that was part of a king's grant three centuries ago is now part of Mel's used-truck lot. Strictly legit.

He got into old trucks (Mel, not the king) because they always fascinated him. He's been all over the country in one of his three cars. He goes on the Interstates sometimes, loving those big powerful trucks growling along, the old Kenworths, Macks, Peterbilts. D'ya know where that name came from? Peterbilt? 'Cause Peter built 'em. Honest. But Interstates are so sterile. Most of them don't even allow billboards since LBJ's missus got on that kick. Mel prefers the slow lane on the back roads like 1, which he has driven up and down several times. Those lanes let you savor life, not zip straight through it, racking up the miles on the fast track to somewhere you won't even remember in a couple years. It's the doing that matters, not the having done. The slow lane is full of local sights and sounds and dotted with deals. Mel buys old signs for his collection, as Jim Bennett does a thousand miles down U.S. 1. But that's Georgia. Mel's got an eighteen-footer for Tone-Up, a defunct soda pop; the sign was probably what was holding up the side of some old wooden barn. Mel brought that one back to Portsmouth from Kansas City. Antique trucks, too: He's got one from 1910. But that baby's not 4 Sale.

Mel got frustrated with selling new cars whose prices can be limited by annoying things like competition and customer knowledge. New cars have stickers right on the window, so you're in the same ballpark the minute that guy hits the showroom floor. And U.S. 1 does not suffer from a dearth of showroom floors. Used cars, though, can be a different story. Every one is different. And so are used-car customers, who've changed. They are nowhere near as knowledgeable anymore. The MTV generation doesn't have the time, or take the time, to shop around. They decide they want something today, this afternoon, this minute, they want it now: clothes, cakes, soups, diets, affluence, highs, boy babies, cars. Isn't that why God invented microwaves, drive-thrus, commodity markets, instant potatoes, genetic engineering, Lean Cuisine, Jiffy Lubes, dead turkeys that baste themselves, and used-car lots with little plastic flags and the neighborhood's highest electric bill? Even He worked on Saturday to get the job done sooner.

The car's been sitting on the lot for six weeks, patiently looking all shiny and nice under those many bare bulbs. But now, suddenly, this couple walks in on the way home from the grocery store, with their rented movie and a bottle of California blanc. They decided to buy during breakfast, after running but before the strawberries. They want their new old car now—ASAP, not next month after a lot of messy haggling, going here and there. So a dealer can name a price. The couple mumbles a little. The dealer lowers it a little. And the couple says, well, okay, fine. Add it to the home-equity loan.

Where's the fun in selling like that?

Today's customers don't like to save a few bucks and fix something themselves, either. Same for today's home buyers. They want something ready to use *now*. Handyman specials are out. The vehicle must be ready to go now. So the dealer has to fix them up as if the piece of junk is new, for Mel's sake. What do they expect for four grand?

Used trucks, however, offer a real opportunity. Who knows what they're worth, right? They're worth what we agree on. Also used buses. Good sales. Especially if you develop those Central American markets that are always having revolutions. Revolutions tend to have a lot of explosions. Coups can, too. Explosions use up a fair number of used trucks and buses, all of which helps the turnover. So there is a steady stream of Central American used-truck and used-bus buyers in tropical slacks strolling into Mel's office inside the chain-link fence in Portsmouth and not speaking English. Which is no problema, amigo, because Mel doesn't speak Spanish.

Mel speaks Deal. Understand? Now, you might ask, how do Spanish-speaking used-truck buyers seeking vehicles to put land mines under come to be in Portsmouth, New Hampshire? By airplane, of course. No, just kidding. Listen, these tired buyers are especially tired of being ripped off by overpriced, unscrupulous used-truck and -bus dealers down near the border with Mexico, which hasn't had a revolution in a few years, but give 'em time, you know what I mean? Mexico is close to those countries that do have regular revolutions. And many Mexican customs officials seem to be former used-car dealers. You hear what I'm saying? In other words, they speak Deal, too. Got it?

So Mel keeps in touch with all the truck-leasing companies up here. He knows what his buyers want— Mercedes are most popular—and he helps these little guys out. Collects some used trucks. Gives them a very good price, unlike those Texans. But it'd be a real shame to take all those trucks all that way empty. So maybe Mel helps them buy a load of washing machines and chain saws and TV's, a few of which might be seized with a smile by some customs officials and the rest of which will end up cleaning the clothes of dirty revolutionaries, cutting down trees for roadblocks, or showing films of the latest unsmiling cabal in perfectly pressed custom-made camouflage suits. Such sideline sales may not show up on the nation's balance of payments reports, but they help finance the whole damn truck-buying trip.

Not your classic mercantilism. But it proves Yankee traders ain't dead. And it's the best kind of deal: Every-body wins. Not too bad for a former seventh-grade used-bike salesman.

Now, gotta go.

In modern times, U.S. 1 has been known in several areas as "Old Bloody" because of the many traffic accidents that occur in and along its lanes, as if it is the road's fault. But like any major thoroughfare, even the predecessor roads to 1 have been the scene of some nefarious doings. To be sure, P. Revere made very good time between Boston and Portsmouth with his revolutionary news. The Hessians marched along the old roadway, which was new then, and the speedy response of the Minutemen (the militia, not the muffler repairmen) would have been impossible without a more than passable road. General Washington's troops used the road when they delivered their Christmas surprise at Trenton. Washington, on his postelection national tours to Georgia and New England, rode its bumpy length in the days before presidents needed Secret Service cavalcades to protect the country's leader from his countrymen. Like Duncan Hines, President Washington kept a detailed diary of his travels and his opinion of each night's inn and its fare, opinions that benefited in candor because innkeepers didn't read a lot of nonledger books and presidents didn't sign lucrative book contracts in those days. (Neither did their valets.)

These were the days before frightened Americans began looking inward so much, before they fled to the basement family room, the fenced backyard, the enclosed mall, and the unlisted phone number. Back then, front porches provided open windows on the wider world, which was the main street. Being so central to local life, the main road was also a public punishment place. In 1662, a Constable Richard Waldron, not known for his religious tolerance, sentenced Anne Colman, Mary Tompkins, and Alice Ambrose to be tied to a cart and pulled by the officer down the road from town to town, "and driving your cart through your several towns, to whip them upon their naked backs not exceeding ten stripes apiece on each of them, in each town" until such time as they were conveyed out of the jurisdiction.

Whittier wrote:

Bared to the waist for north wind's grip,
And keener sting of the constable's whip,
The blood that followed each hissing blow
Froze as it sprinkled the winter snow.

Such penalties do not fit the self-image of warm hospitality in New England, where Puritan ministers believed that the slightest wrongdoing by an individual could wreak havoc on the entire community. Therefore, each punishment was obviously warranted for "vagabond Quakers" who had wandered into an outpost of the Massachusetts Bay Colony. Such punishments also created commercial possibilities by attracting sizable roadside Colonial crowds, which the Hampton Beach Casino wouldn't mind today. The attraction is on U.S. 1A, which is what entrepreneurs in New Hampshire and Florida call another parallel road closer to the ocean so they can seem to have two main streets for attractions, fast-food joints, and motels. The Hampton Beach Casino offers seven acres of fun, forty major attractions, parking for seven hundred cars, and "big-name entertainment" to get those all-important tourists off the Interstate. The theme park and elaborate water slides are now open every day, April through October, because for the last 175 years or so, traveling on Sunday has been permitted in New Hampshire.

New Hampshire is the state with the license plates that proclaim "Live Free or Die," the kind of public slogan that caused people to nod patriotically back in the fifties but now, with less clear-cut enemies loose in the world, prompts some people to look sideways at each other without saying a word. But what politician could ever propose erasing such a subtle thought from something as vital to human discourse as a license plate?

Hampton was originally called Winnicummet ("beau-tiful place in the pines") by the Indians, who thought it was. But they didn't make it to the town meeting in 1639 when the pioneers' minds and votes turned back toward the British homeland. So Rand McNally has gone with Hampton. Hampton's white settlers planted lots of elms and a few shoe factories. They all withered eventually, although some of the original pines lasted long enough to soak up some acid rain.

Apparently, the devil really liked U.S. 1 in New Hampshire and nearby Massachusetts, because their histories are full of stories concerning evil spirits. Goody Cole, one of New England's all-time notorious witches, had her hut just off the roadside in Hampton. It was known, not proven but very well known nonetheless, that she had made "a league" with the devil, who doesn't get his name capitalized. And this alliance enabled her to deform people, to torture them, and even to drown them without her being anywhere in the vicinity, which explains a lot of things over all these years. After several years of indeterminate imprisonment, Goody asked Justice Jonas Clark if she might be let out. He said no, because while she was "not Legally guilty" there was "just ground of vehement suspissyon of her haveing had famillyarrty with the deiull."

Hampton was a prominent stagecoach stop on the Post Road to Boston and also the home of Jonathan Moulton, a Revolutionary War general (was *everyone* a general in that conflict?). According to local legend, Moulton was famed for his frugality, probably due to his removing the rings of his dead first wife and presenting them to his second spouse. No. 1 Mrs. Moulton is said to have been annoyed, so she forever haunted the place. Then, General Moulton is said to have done a Faustian deal with the devil to obtain all the gold that his boots could hold. The general placed his boots in the fireplace. (Apparently, without the money-laundering marvel of electronic transfers, devils back then preferred to make deliveries via the chimney, like Santa Claus.) But tricky little human devil that he was, the general cut off the toes of his boots to assure a larger flow of gold. This not only got the general in dutch with the devil, it cost him one very good pair of boots.

Not far off U.S. 1 on a rural sideroad, there is also said to be a large stone buried deep in moss back in the woods. This stone, according to those who heard the story from some who knew the people who swore they once heard it from a minister who said he saw them, carried strange chiseled inscriptions. Some dismissed the marks as early trailside graffiti left by bored Indian teenagers. Others seriously suggested that the Norsemen might indeed have crossed the King's Road well before it became U.S. 1. (One can easily imagine the coverage the marked stones might have drawn in Pilgrim tabloids: PAGAN ALIENS LAND IN WOODS / Hairy, Horned Men Eat, Sing While Waving Swords.)

Another Town Named ——port

SALISBURY IS THE FIRST community that travelers on U.S. 1 encounter in Massachusetts, if they're keeping track of states, which isn't as important as it was when visitors with the wrong religious beliefs could run into terminal difficulties right here on earth without waiting for the hereafter. Even death could be dangerous in that wild northeastern corner of Massachusetts in the early days of white settlement. So men and women of affluence had large stone slabs placed atop their grave to avoid being dug up by hungry wolves.

For many years Salisbury (yes, the renaming-from-England bit again) also preserved the old Whipping Stone that the Quakers used to hold down troublesome live people during proper punishment. Then one day, perhaps during a membership drive on the advice of a public relations consultant who saw some contradiction between flogging and a religious group calling itself the Society of Friends, the Quakers broke their whipping habit and got into prison reform. Reform-minded Quakers, at least the branch down the highway in Pennsylvania, invented the penitentiary. It was intended to provide decent prison living conditions so that the guilty could be penitent, ponder their wrongs, and learn not to do them anymore, lest they be punished again and have their reputation further tarnished in this compact little country.

The penitentiary, like many reforms, seemed like a real good idea at the time. It had worked with some youngsters sent to their rooms without dinner to ponder their wrong. As American social reforms go, it was a major change for prisons, which had been mainly warehouses to store the guilty briefly (at their own expense) until trial or administration of the real punishment—physical abuse, exile, or execution. But this prison reform didn't work, either, Americans not really being into penitence, which can take a lot of time, you know, if you're not a hungry youngster. And while chopping off the right hand of a pickpocket certainly seemed decisive to the voters and certainly eliminated one right-handed pickpocket, it didn't seem to do much for the larger problem of theft. Today's popular solution, after two hundred years of experience, is to build more prisons and keep the inmates so busy that they won't have time to plot new crimes. Idle minds make mischief, you know. I'm sure that will work.

Newburyport is the smallest city in Massachusetts, and one of the prettiest, which may help the new hotel attract outsiders. The city's population has hovered around the same 17,000 for a century (not really the same people but the same total). John P. Marquand, the novelist who lived in Newburyport awhile, said it "is not a museum piece although it sometimes looks it." Newburyport was also known as Yankee City in the famous social studies of American city life that focused there in the 1930s and '40s.

Originally built to serve the seafaring trade, Newburyport was the founding site of the U.S. Coast Guard and home to captains and traders and silversmiths (whose corporate descendants still work there) and wily merchants like the self-appointed "Lord" Timothy Dexter. He made a bundle speculating in Colonial currency, selling bed-warming pans, and could have made more if he'd had access to late-night television. He died in 1806, so he is not blamed for the shifting tidal sands that silted up the nearby mouth of the Merrimack River in the late 1800s and limited the size of oceangoing ships that could dock at Newburyport.

His ghost, however, could have had a hand in the 1877 fire that wiped out the downtown. Devastated, town fathers debated how to rebuild as economically as possible. The solution: Use all the bricks that the ships had carried home as ballast before simply dumping them on the city's riverbanks.

By 1960, however, the community was hit with the same kind of urban blight that had seen many cities' citizens seek their future elsewhere, leaving the downtown without a present, boarded up, bypassed by traffic and hopes. One proposal was to bulldoze everything and start over. But wait, said some. What if we kept the character and rebuilt from within? Newburyport became Massachusetts's first city to get Federal urban renewal money up U.S. 1.

Of course, being America and New England, doing anything required extensive debate at evening meetings. Also quorum challenges. And committees, lots of them. This has gone on for nearly thirty years. "We move slowly here," says a smiling Janet Richey, who should know. "No one ever says, 'This is a great idea!' And the crowd rises up as one to agree. We've got to chew on it, long and carefully, every tiny piece of it."

Miss Richey has lived in Newburyport for only two decades, but already she has become head of the Newburyport Redevelopment Authority, a job that is now called chairperson. This is in addition to making homemade fudge at her own candy store. She was a certified public accountant for many years in a large city, but like many city dwellers in recent times, she grew tired of the crowds, the hassles, and the fears of big-city life that can accompany an apartment buzzer, a shadowy hallway, or an empty elevator. (One popular panacea: Put a large mirror in the elevator so someone paying hundreds of dollars a month in rent can at least see a mugger before he strikes.) "I knew, at least I hoped, there had to be another way," recalls Miss Richey. She looked around for a modest-sized city that was large enough to have a downtown, small enough to know the regulars who gather at the same café every morning, including the mayor, and close enough to other cities to permit easy non-rush-hour trips on modern highways to cultural diversions. She tried California for a year, and Colorado. Newburyport, less than a half hour from Portsmouth's reasonable restaurants and under an hour from the Boston Pops, won her lease.

She rented an apartment over a small store at 21 Market Square and opened Something Sweet. "It seemed like a good idea." And if that has seen her go from a size eight to a size ten, it also has brought joys and satisfactions and friendships, not to mention a valued proximity to the ocean. Miss Richey does some work seven days a week, but it's her work. She has five part-time employees and a growing array of candy recipes, which she merges and mingles and tests and sells. She's noticed that people's tastes, like the Merrimack's sandbars, are shifting in recent years. People are always seeking something different now, radical, even bizarre in language, dress, entertainment, even candies—not so much the plain chocolates anymore, but more exotic fare. Chocolate-covered apricots are big. The success of chocolate-covered peanuts didn't surprise her. But chocolate-laden pretzels? "I thought, What good Yankee would go for that?" It's a big seller.

Part of that is due to the changing people. A fair

portion of the population scattered along Newburyport's tidy streets and the nine square (and very flat) miles of incorporated countryside are like Miss Richey; they come from somewhere else, which produces an occasional clash of values in many communities. Typically, the newcomers want to keep things the way they are. Old-timers wouldn't mind that, if they could make a living, which doesn't seem possible after some years of despair. Unlike some new suburbs, Newburyport's newcomers seem to develop a personal stake in the community. Many still commute to Boston and Peabody, places like that. But they seem less likely now to pack up and transfer to the next similar suburb whenever the company says "Jump." For one thing, their spouse typically has a job, too, doubling the occupational obstacles to negotiate. And the natives seem silently more appreciative of the new blood, or at least accepting.

So while it may take two generations, Newburyport's conflicting interests have compromised and made considerable progress toward restoring the physical community while adapting to new needs. Threadneedle Alley becomes Harris Street and Fish Street becomes State Street. The buildings downtown look much like a prosperous Newburyport of yesteryear (with indoor plumbing). The signs are old-fashioned, but there is a small industrial park hidden on the edge of town, producing electronic parts and stains, among other things. And the community is going after that prized pollution-free resource—tourists.

In fact, that's what the latest fighting was about, building a small convention center as part of a 120-room hotel that would be the jewel of Newburyport's reborn waterfront. The idea is simultaneously to attract business people and shoppers, with special weekend getaway rates for city couples and even seasonal attractions for families, with the springtime teddy bear picnic and the candlelight Santa parades. However, building such a modern hostelry requires numerous zoning variances, which, once proposed and/or granted, are inevitably challenged, with the losing side appealing through several layers of courts, as far as its money will pay its lawyers to go. And then

because the convention center will be on the water, some special state permissions are necessary—and challengeable. And they need to guarantee the general public's access to the water without depleting the location's peacefulness. And then there is the convention-center developer, who is less interested in retaining lawyers ad infinitum than in making money, and maybe his patience seems to have less staying power than all these challenges. So maybe another variance or property-tax sweetener is necessary to assuage him and that *is* challengeable. And so on. And so on. And so on. But there is an endless supply of evenings to meet on, and lawyers to bring bulging briefcases along. Miss Richey is as patient as a chairperson needs to be in a modern American community accommodating newly diverse interests. And if things get a little too frustrating after some sessions, she can go back to her apartment, change clothes, and pound fudge for a few hours.

Time was when communities such as Newburyport along the highway had to be pretty much self-contained. The only way out in Colonial times was to buy a horse, ride it out, and then sell it at your destination, which is why used-horse traders were so abundant. Private ownership of anything as grand as a carriage was considered pretentious in 1700. And so there soon after developed the stagecoach business, bumping buggies that bounced along the trails irregularly, stopping and starting according to the whims, thirst, and sobriety of drivers who expected even frilled passengers to push when mud claimed a wheel.

Probably the country's earliest intercity land transportation was the wagons that in the 1630s began hauling freight and an occasional hardy human from Trenton, New Jersey, northeast to the port of Perth Amboy for transfer to ships for the last leg to New York City. Until then, and even for many long years afterward, the waterways were the main mode of transport. The main Post Road (aka King's Highway, Boston Post Road, U.S. 1) didn't become the nation's initial cement tendon until the

1930s, after the assembly line and the automobile had democratized American travel. But in 1781, the new nation's main street was the main Post Road from Wiscasset, Maine, all the way down to coastal Georgia. In that decade, fully fifty-one of the country's two hundred post offices were located on the same road (many still are). So important was this single highway to American life that other roads leading off of it were officially designated and popularly referred to as *Crossroads,* a term later expanded to its looser usage of any intersecting roads anywhere. But originally, all crossroads led to 1.

The first successful regular New England staging business was the Portsmouth Flying Coach, launched in 1763 with great optimism and some success by John Stavers, despite his suspected Tory sympathies. *Flying* meant the coach was suspended on straps, presumably for a somewhat smoother ride. *Flying* was also eighteenth-century advertising hyperbole for six miles an hour. It took Mr. Stavers and his speedy relay teams at least two days to go from Boston through Saugus and Danvers, where the malls are now, to Newburyport and up past the soon-to-be-haunted, soon-to-be-General Moulton home to Portsmouth. The Stavers stage ran each way once a week, leaving Boston's Charles Street ferry early Thursday, first stop Saugus, which was the birthplace of America's steel industry because of the bog ore discovered there. This discovery lead to numerous factories turning out kettles, anchors, dies for coining America's first coins, and wrought iron for future antiques.

Saugus is now more famed as the home of an important restaurant, part of the nation's burgeoning service industries. Route 1 restaurants usually uphold the age-old American commercial dictum: Not just good food fast but fast food quick (and never mind what the customer actually ordered). But then there is the Hilltop Steak House, an intimate 70,000-square-foot restaurant that seats 1,400 diners in a western kitsch decor of wooden Indians and deer antlers and proves that in modern America, more is better. At least in Saugus. Every week, Hilltop's diners wash down a small herd of cattle

(40,000 pounds!) with, among other libations, one hundred cases of Budweiser. So successful is the well-organized Hilltop in packing eaters into an area nearly the size of two football fields that, even with prices below more fashionable establishments, it regularly ranks number one in dollar volume among the nation's one hundred largest independent restaurants. Since the mid-fifties, the secret of founder Frank Giuffrida has been simple: reasonable prices for more good food than nearly anyone can eat (so much so that the Hilltop spends about a thousand dollars a month just on that wonderful American institution, the doggie bags). And the herd of plastic cattle out front with the sixty-eight-foot-high neon cactus doesn't hurt traffic at all, either.

The next Stavers stage stop was Danvers, which under the name of Salem village in the late 1600s was the scene of so much witch hysteria that twenty people had to be executed to make it right. This was the perfectly Puritan, supremely somber setting for *The Scarlet Letter,* Nathaniel Hawthorne's biggest seller. (Salem was also his hometown.) Danvers presently marks the unofficial northern end of that vast metropolitan sprawl that lines hundreds of miles of the Eastern Seaboard along the spine of U.S. 1 down through New York-Newark-Philadelphia-Baltimore-Washington-Richmond and on into central Virginia.

The Stavers stage may have detoured slightly at times onto the future U.S. 1A into Ipswich to hit the Whipple House (an early package store, one quart minimum and selling "to any but men of family and of good repute"). By late Friday, the stage made it to Portsmouth's Earl of Halifax Inn, a fine establishment that saw fit to change its name to Pitt Tavern at the suggestion of a Patriot mob that had made its statement by shattering every window in the building. By 1773, Mr. Ezra Lunt opened a competing stage business, but he soon sold out to join the Continental army. And the stagecoach business went into a brief period of decline, owing to the military unpleasantness abroad in the land.

The price of the ride in those days was thirteen shillings per person. Today, it costs $4.65 each way ($133 for a monthly pass) for a bus ride, which begins in Exeter, New Hampshire, stops in Newburyport, and an hour later, traffic permitting, which it often doesn't, arrives in downtown Boston. The stage was open to the air. The bus is air-conditioned. The stage carried perhaps six passengers. The bus holds forty-seven and has put on extra vehicles in recent times as the rising price of fuel prompts more commuters to leave the driving to others. And when too many callers telephone the bus company at the same time and the recording on the automatic answering machine sincerely apologizes for the delay, the machine also thoughtfully pipes easy-listening radio music into the ears of each caller put on hold. (Business psychologists find this helps hold potential customers, who are very impatient.)

In the nation's youth, Boston and Philadelphia were the major centers of travel by stagecoach, which played a crucial political, psychological, and even cultural role in developing the country's sense of identity. By 1784, a person with the need to travel, a small sum of money, and a hardy bottom could ride stages from Portsmouth, New Hampshire, all the way down to Richmond, Virginia. Two years later, stage lines reached down into Georgia. By 1802, regular stages connected Boston and Savannah, Georgia, a distance of some 1,200 miles, averaging fifty-three miles a day. The three-week excursion fare was $70, though you had to spend three Saturdays on the road. The ensuing decades would see branch lines move out on the nation's major roadway north and south from Portsmouth up to Portland past Marm Freeman's Cape Neddick House, which burned down one day when the pot got too hot. From Portland, Colonel T. S. Estabrook eventually operated daily wagons along the coast through Brunswick, Bath, Wiscasset, Belfast, and on up to Montreal. Down South, the stages ran less frequently, a measure of the less-dense population that survives to this day, and of the expense of hauling hay and other supplies into the isolated stage stops that dotted the sandy barrens of southeastern routes.

By 1810, scores of stages left Boston every day, each from its own favorite inn. Philadelphia had thirty-eight stage lines with 213 departures weekly. Ten years later, there were seventy lines and 520 weekly departures. Some historians have calculated, with a certainty that can derive only from the knowledge that no one else is counting, that in 1801, stages covered 24,490 miles every week. Many went north and south along the route of 1; others headed west for the long haul to the frontier. Such regular departures of transportation, at times every twenty minutes, enabled the development of other population centers close by. These are now called suburbs. But such individualized communities would not be possible until the perfection of individualized transportation.

But, of course, Levi Pease, an early stagecoach operator running out of Boston toward New York, could have known nothing of that when he boasted in a 1786 newspaper advertisement: "By the present regulation of the stages, it is certainly the most convenient and expeditious way of traveling that can possibly be had in America, and in order to make it the cheapest, the proprietors of the stages have lowered their price from fourpence to threepence a mile, with liberty to passengers to carry fourteen pounds baggage."

The condition of all roads was left to each local township, which made them just fine as far as the edge of town. The few travelers going beyond those bounds were on their own. That was acceptable so long as the country remained an unconnected collection of self-sufficient local communities, coincidentally adrift on the same continental land mass. The diaries of early travelers refer to execrable road conditions. That situation persisted for another century until after a 1919 motorized military expedition under the command of a young army captain was ordered to attempt an experimental cross-country land journey on the nation's back roads (there being none in the front at that time). At times, that modern military convoy could go no faster than old-time stagecoaches.

It was no coincidence that within five years of the captain's report, American highways were being systematized and paved and numbered, beginning with 1, which was called "the most important road, everything considered." Nor was it a coincidence that thirty-five years later that same army captain would launch construction of the world's largest transcontinental network of superhighways, a cement legacy that would change the face of the country—and many cities. Commanding the greatest military invasion in history and conquering the Nazis with their own paved network of national expressways had taught Dwight Eisenhower many things about power, but none more vivid than the lessons he had learned struggling along some simple muddy roads.

By early-nineteenth-century standards, the stage roads were slowly improved. Privately owned turnpikes played a major role since there was money to be made by offering a smoother ride, even if that "smoother" ride was over tooth-jarring miles of irregular logs laid side by side, corduroy style. With the imminent arrival of the railroad, which created the concept of mass transportation, and the telegraph, which negated the need for much travel by messengers, such road improvements would pause for a few decades until reignited by the pressures of massive individualized transportation, the private automobile.

Nonetheless, those early road improvements did allow more direct routes, enabling express stagecoaches to cover upwards of ten miles in an hour, farther than cars go on many long parts of the crowded road today. Before 1800, the postal service was offering four-day letter service between New York and Boston, speedy delivery that is sometimes matched today. By 1827, the eve of the stagecoach's twenty-year heyday, travel time between New York and Philadelphia had been cut from twenty-three to sixteen hours. Today, thanks to the introduction of Zip codes, airplanes, computers, trucks, and high-speed sorters, the overburdened postal service requires at least forty-eight hours.

With a start, some European visitors experienced through their travels by stage some amazing aspects of American democracy at work. They had to get their own food from the buffet table, for instance. While supping, some drivers actually sat next to persons of consequence! And even stage drivers addressed the most noble of passengers in an open and frank manner far and above their traditional station. Acting as if you owned the place was fine if you did. But such an attitude has created a fair number of cultural misunderstandings in more recent times when Americans carried it abroad. In such a studiously egalitarian atmosphere as an old inn, which was called an "ordinary," it was considered demeaning to offer a tip to servants. This was one aspect of democracy the foreigners could become accustomed to quickly, and one that many Americans ultimately could get over. (The European visitors also noted, with not a little wonder, that American travelers tended to shun the whip in favor of talking to their horses, as if the animals had a personality. This unusual puzzling penchant for attributing human qualities to animals, seen later in the lingering popularity of everything from *Moby Dick* to Mickey Mouse, Lassie, Rin Tin Tin, and even Mr. Ed, may help explain American opposition to whale hunting, seal bashing, and more recently, fur trapping. It doesn't, however, surprise veterinarians like Larry Berkwitt. But that's a story for farther down U.S. 1.)

Stagecoach drivers were impressive and romantic figures sitting high atop their fast-moving vehicles, urging their steadfast steeds to greater speeds, moving from one place to another as free as the wind, and paid to do it. Dashing men of the world in a nation of stay-at-homes. Drunken driving could be a problem even then, with unthinking passengers often offering to buy their designated driver a round at each tavern stop, the early equivalent of today's truck stops. Some companies required drivers to sign pledges "to abstain from the use of Ardent Spirits as a Drink while Engaged in Running Stages."

Approaching each community from afar, these drivers took the bouncing reins in one hand and with the other sounded their brass horns to alert the innkeepers, one blast for each passenger requiring a meal. The horn also summoned residents to hear the latest gossip from far away and perhaps to receive a letter. Even if there was no mail addressed to them, townspeople thought nothing of opening letters addressed to others in order to see what was going on. This prompted development of private codes between correspondents to protect their privacy. This habit, plus the advent of highwaymen in rural parts of Pennsylvania, also helps explain the reluctance of people to send money in the mail and the popularity of letters of credit.

Under contract with the government, stages carried the federal mails between large communities. But the fee for whatever local missives, packages, or even an oxen or two that the drivers picked up and dropped off en route supplemented their substantial wages, sometimes approaching fourteen dollars a month.

At first, every American newspaper could send one copy of each issue free to every other newspaper. This was designed to encourage the spread of ideas and democracy. It did help spread literacy—also plagiarism. After some years, the post office began charging for mailed newspapers, which prompted heated denunciations of this "news tax" by the likes of James Madison. It also prompted publishers to do their own deals with stage operators: You haul the papers for free and I'll run your ads. Watching one literate citizen read a newly arrived newspaper aloud to his friends prompted one Englishman to note in his diary, "The knowledge acquired by newspapers may be superficial, but it gives men a general acquaintance with the world." And that was before colored charts!

On some stretches of straight highway, some stages might make upwards of ninety to one hundred miles a day. To change horses, they stopped at the same inn every trip, a privilege for which even old-time innkeepers might well have paid a little something. Delayed by weather or whim, the stages might arrive for an overnight stop around midnight. Warned that departure would come at first light, weary travelers would fall into their inn beds of straw after a hasty meal. Soon, they might be awakened by a candle-carrying innkeeper ordering them to move over for a stranger just arrived on some other tardy stage. Then at 4:30 or so, the driver, who did not own a Timex, might well be shouting in the darkness that departure would be in thirty minutes.

My family's trips also began before dawn, although why anyone was worried about beating the traffic at 5:30 on a Saturday morning late in the 1940s always escaped me. With armloads of luggage, Dad and I would make trip after trip out the back-porch door to the car while Mom busied herself with shutting down the house for an entire week, which was considered a long vacation in my family. She also packed the first day's lunch in a metal picnic basket to be opened at one of the hundreds of weathered public picnic tables that dotted the exotic roadsides everywhere. It was not only more economical to take your own lunch in those pre-Interstate days, it was safer. Each town had its own restaurant, but being travelers, we never knew about the quality. Until Mom met Duncan Hines. And, according to my parents' rigid rule, we couldn't eat there if the lighted neon script had the word *Liquor* larger than the words *Good Food*.

I was never bothered horribly about the quality of restaurant food on the road, since there isn't too much gustatory violence anyone can do to a peanut butter and jelly sandwich, and that is what I ate in those days. Since I had found the perfect food (but you can't eat popcorn for lunch around your own parents), I saw no need to test the culinary waters beyond No. 2, PB&J. Those were the days before so many national chains, of course, when it was assumed that a restaurant named, say, Bob & Betty's was more than just a registered corporate logo. It was not naïve then to casually assume that such a place had something to do with a couple named Bob and Betty. Even Howard Johnson's, no doubt, had an owner named Howard Johnson, who liked orange roofs and might walk out of the kitchen at any moment, having just finished whipping up a batch of yet another obscure ice cream flavor. Mr. Johnson certainly knew the secret to toasting and then buttering a hot dog bun the way it happened only under his orange roof. I do not know for a fact, but I suspect that Roy Rogers has never been seen wearing a plastic tag that says "Howdy! My Name is ———" anywhere near a french-fryer at places that carry his famous name. (But then I always liked Gene Autry better, anyway.)

For many minutes after being seated in one of those roadside places, my body still felt as if it was thumping over the road's repetitive tar lines as it had been all day. The roadside lunch ("Andrew, use your napkin, please") had been relaxing and probably followed by a short snooze on the ground on a heavy woolen steamer blanket. I was puzzled early on because I never saw steamer blankets steaming, not even in winter when I could see my breath. My Grandmother M. informed me steamer blankets got their name because people wrapped them around their legs while relaxing in fold-up wooden chairs on the stern of immense oceaner steamers, which were not yet glamorous and steeped in tropical sunshine. (Because she was a schoolteacher renowned for knocking knuckles with wicked wooden rulers, I never did ask her why anyone in their right mind would sit out on a salt-sticky deck in weather that required wrapping limbs in heavy blankets made of used sheep hair.) Now, these coverings have a different name and are lightweight because the astronauts developed them for freezing football fans who can't carry heavy things that aren't liquid all the way up to their plastic seats. Hence, the name stadium blanket. Anyway, our steamer blanket came from the car trunk and it did feel warm, even in the shade and even if it wasn't acrylic.

But now in this strange restaurant it was dinnertime. And soon it would be the proper hour to pull in the driveway of a tourist home that was always run by somebody's grandparents. We didn't stay in motels in those days because they had a reputation for having "hot pillows" and hourly rates, which adults knew was bad because J. Edgar Hoover had said so. We might stay in a tourist court with the circular gravel lane and the wooden chairs in front of each cabin and the beds that squeaked differently than the ones at home.

Thirty years before my childhood, the early days of pleasure motoring had been marked by unpleasant breakdowns and not always comfortable campouts in fields. So risky was car travel then that motorists went in caravans. Road signs were nonexistent, as was heavy traffic. So motoring clubs erected their own road signs or had members toss fistfuls of shredded paper out the window to ease their return trip. By 1925, Socony, the corporate predecessor to Mobil, was launching yet another hopeful public-service battle over America's mass manners, which has yet to be won seven decades later. Socony added dual warnings to its free road maps: "Don't Leave Smouldering Camp Fires! Paper thrown along highways is unsightly."

The top speed of America's first police car, a Stanley Steamer that began patrolling U.S. 1 and the other streets of Boston in 1903, was ten miles an hour. Until then, police desirous of halting speeders (wreckless souls who went over eight miles an hour) had two choices: shoot the speeders or jump on police bicycles and give chase, images that now seem suitable for a silent movie. In 1895, just a few years after the Duryeas built their successful gasoline-powered car, promoters staged the first car race, a grueling fifty-mile course that winner Frank Duryea covered in the blazing time of ten hours. That same year, Thomas Edison had made a prediction that drew considerable news coverage. "The horseless vehicle is the coming wonder," he said. "It is only a question of time when the cars and trucks in every large city will be run by motors."

In 1900, *The Saturday Evening Post* published the first automobile advertisement. By 1910, 458,500 motor vehicles had been sold, although only twelve states yet required any kind of license. (New York, of course, was the first to see the revenue potential in cars, first, requiring motor-vehicle-registration tags and then offering them for sale. This law brought in $954 during its first year, 1901.) The year 1910 also saw the invention of the coffee break at the Springfield auto plant of Rolls-Royce, a twice-daily institution invented to accomplish two of the plant managers' goals, neither of which had to do with promoting socialization among employees. Having an office boy carry the hot beverage to workers kept them on the assembly line and it negated the need to install an expensive cafeteria.

During their breaks, Americans were talking more about the unmanageable ills of their crowded cities, and Henry Ford, perhaps with just a tad bit of understandable self-interest, had a good idea: "We shall solve the city problem," he said, "by leaving the city." Guess how he suggested that departure be made. And guess what invention also allowed the far-reaching development of the suburb, which would change for better or worse for many generations so many aspects of American life, not to mention the balance of payments and the traffic conditions down in places like U.S. 1 in New Jersey.

By 1917, America had 5 million motor vehicles on the road. Woodrow Wilson, then, became the last president to ride to his inauguration in a horse-drawn carriage. In 1920, the first flat tire was discovered, an inevitable side effect of the change from solid rubber wheels to smoother-riding inflatable ones. By 1925, building good roads had joined supporting motherhood and opposing Bolshevism as the most popular political platforms; today there are 3.9 million miles of roads, 90 percent of them paved, and Americans think nothing of leaving road building and maintenance to government, a job that once belonged to civic-minded citizens or private business.

Just such a company began constructing the Newburyport Turnpike to Boston in 1803 in an early effort to shorten by ten miles the route along the winding forty-mile coastal path, thereby saving time and money. The laborers, who provided their own pickaxes and shovels, earned under a dollar a day (more if they provided their own oxen). One contractor received eight thousand dollars and a hogshead of rum for the work on his 3.5-mile section of the future U.S. 1. The laborers built more than a hundred bridges, whacked perhaps twenty feet off the top of hills, and regraded and widened the roadway so tree branches wouldn't scrape the horse teams. As travel speeds increased, it became more important to denude the roadside. So the old elms and oaks had to go.

Turnpike tolls were collected at three booths along the way, five cents for a man on horseback, ten cents for a two-wheeled vehicle, and a quarter for four-horse coaches. Pigs and sheep being driven by shepherds were three cents per dozen, but pedestrians were free. In the War of 1812, many troops rushed up and down the road, trying to anticipate where on Massachusetts's 1,500 miles of coast the British would raid next. During the Civil War, thousands of local men and boys marched away along the turnpike (for free), many to pay the ultimate price elsewhere. Over four decades later, the roadside provided training grounds for troops for the Spanish-American War and the Great War, which is what World War I was called until we got a second one.

Without telephones and televisions to socialize at home, travel in those days provided a sociable experience with much stopping along the way to chat and gossip. Even without an 800 reservations number, the turnpike company built its own hotels and inns to serve travelers. Over the years, many roadside service businesses sprang up, everything from livery stables to diners, from inns to motels that did not demand to see a marriage license, from highwaymen to repairmen who lived off the wrecks and misfortunes of machines the way some coastal residents lived off maritime salvage.

It took two years to build the entire Newburyport Turnpike, less time than it takes to repair smaller sections today. And it cost a half-million dollars, half the price *per mile* of turnpike construction 150 years later. By changing horses halfway, some accounts say the flying mail coaches could accomplish a round-trip to Boston in four hours, for $2.50. So brisk was demand that the Eastern Stage Company, a corporate successor to old Mr. Stavers, maintained thirty-five coaches and nearly three hundred horses. Today, the Boston-Newburyport local highway has but one surviving Horse Xing sign, and the coach company runs just seven buses along the route. But the lumbering vehicles are available for charter by any group, especially the easygoing autumnal crowds of "leaf-peepers."

Despite its speed and modern imperative to make time, the Newburyport Turnpike never did make money for its investors. Too many travelers cheated, getting on the turnpike after one toll booth and getting off before the next, which led to the invention of the guardrail. By the mid-nineteenth century, along came the railroad, which was definitely faster and smokier. And, some historical records show, a fair number of travelers found the turnpike's hurtling coaches, dashing horses, and narrow lanes too frenetic. So they returned to the slower lane of the old local road.

That sense of hurried disruption continued into more modern times. Nearly a half-century ago, Nellie Burnham Allen of the Committee on Publication of the Danvers Historical Society got to thinking, in print: "One cannot but wonder if and when air traffic has become as congested as road traffic is at the present time, mankind will have developed into a hydra-headed race of monsters with little use of their legs but with the ability to look up, down, forwards, backwards, and sideways at the same time."

Speed has definitely been very important to Americans, the speed of everything and the size of many things from their home state to their television screen to their car and even to parts of the male and female anatomy. By 1925, cars were getting larger—there were 17 million motor vehicles by then—and in just eight years Americans had more than tripled the number of highway fatalities to 22,000 a year. As late as 1940, tuberculosis killed twice as many Americans as car crashes. But by the 1990s, we got that annual traffic total up around 45,000, with half these fatal accidents somehow involving alcohol. Even motor-vehicle accidents became big business; today, they cause $24 billion a year in property damage.

By 1935, Massachusetts had set a maximum speed of thirty-five miles an hour on the open road (being wild and unsophisticated, Connecticut and Vermont still had no speed limits). Connecticut and Maine, however, had, passed laws prohibiting parking any car without also locking it (and Oklahoma City, sensing an opportunity to make some money, installed the first parking meters). Maine, New Hampshire, and Rhode Island did not yet require hand signals when turning. Since bumper stickers had not yet been invented, Connecticut, Massachusetts, and Rhode Island did allow windshield stickers. In the interests of protecting children, drivers were also warned, "Slow down before passing school buses that have stopped to take on or let off passengers."

The first cross-country motoring vacations occurred when a caravan of horseless-carriage enthusiasts successfully wandered far from any cities in 1904, the year when a new Oldsmobile cost $650. Having seen the future at the turn of the century and not liking it one bit, the community of Mitchell, South Dakota, banned all motor vehicles. This commonsense approach in the future hometown of George McGovern was not widely copied. The year 1927 was an important one in the history of the American automobile and highway; that was the first year that cars carried chrome trim. Two years later, the first car radio was installed, though some cities banned them as unsafe distractions. The date of the first in-car conception is not recorded, but some sociology studies years ago produced the believable estimate that 40 percent of all marriages were proposed in cars.

Very quickly, what had begun life as an odd toy for the idle moments of the affluent became an apparent necessity for just about everyone except many New Yorkers, who seemed to prefer stealing or stripping cars to actually owning them. Other material viruses would spread widely through the country in the twentieth century. The telephone, the radio, the airplane, talking movies, television, VCRs, and other inventions all would have their considerable effects on American life and culture, but none any more than the automobile and the highway. Together, they would change the physical shape of the countryside and the cities, as well as the mental framework of much of North American life, outside New York City.

The modern method of mass-producing cars may have been copied in reverse from the disassembly line of animal slaughterhouses. But that process was also adapted to countless other industries, as was the generous pay, which created the financial foundation for the modern

houses. In the United States, you built your own road for the first 150 years. Public bathhouses were out and individual indoor plumbing was in. If you don't like the local rules or conditions, you can always go somewhere else, whether you're Anne Hutchinson seeking religious liberty or a laid-off Massachusetts mill worker seeking work. There has been plenty of room for fresh starts in the United States. And waiting for group solutions can take a very long while here; residents of the new postwar suburbs that exploded around metropolitan areas still would be standing at the bus stop if they waited for their new communities to plan and implement mass-transit operations. Cars have been part of the modern democratic ethic, implying that anyone can go anywhere anytime, even off the public roads by using four-wheel drive. So we bought our own car, on time. And the politicians, who wanted our votes, gave us our roads over time, and hid the costs in gasoline taxes. So my parents and I, along with millions of other identical groupings over the years, went on our annual trips together, gathering the many shared memories. These mental images are so invisibly important in families that first Kodak, then Polaroid, and then all the VCR makers have developed an entire industry to help collect such instantly reviewable photo keepsakes.

Today, commercial jets and excursion fares enable many families to cover more ground in less time, although family members must sit in different rows and everyone listens to a different channel on their plastic headset. I remember my family's car trips as very special. One of the required journeys was the Revolutionary War Tour, which might be merged with the New York City Visit but was definitely distinct from the Civil War Circuit or the College-Visit Trip.

Whatever the destination, Dad did the driving, Mom the navigating, and me the watching. Dad's left arm would get tanned, Mom's right. Dad wore short sleeves on these trips and the back of his shirt always got wrinkled, as did the bottom of Mom's skirt (women didn't wear pants as much then unless they were building bombers). Mom also controlled the radio tuning knob, the buttons becoming useless fifty miles from home and the radio being so old-fashioned that it could not Seek or Scan by itself. Now, thank goodness, the engineers at Matsushita Electric are marketing a new device, something called the CQ-ID90 and ID Logic Circuitry. This is a car radio containing a little computer chip that remembers the frequency of more than ten thousand North American radio stations. You tell the little chip where you live. Then every thirty miles or so, you punch in what direction you are driving. That way, if you don't change direction and you push the radio button for Rock, Jazz, or Clas, the chip will scan its memory quicker than you can say "waste of time." And up on the Dolby speakers will come the Rock, Jazz, or Clas station in that area. No wasted motion. No waste of static. No need to wander the airwaves. Saves time, too—unless, of course, the station has changed its format and neglected to inform your chip.

On our family jaunts, we didn't use the car radio much. In fact, fleeing the usual distractions was part of the vacation, each of us in that common space sharing the same time. Each of us had the other two family members all to ourselves. Those were the days when we had the time to make our own travel discoveries, a time when highway serendipity belonged to everyone, not just Charles Kuralt. The whine of the tires. The whoosh of cars passing in the opposite direction. The air-horn blasts as truckers responded to my hand signals. I sat in the backseat, straddling the huge hot hump in the floor, unless I was discovering the principle of aerodynamic lift holding my flattened hand out the window. I rested my chin on the top of the front seat, keeping a sharp ear for interesting parental observations and an eagle eye out for loose Life Savers and passing curiosities. Mentally, I aimed the large hood ornament down the exact middle of the road. Today, those ornaments are mere trademark designs or wild animals. Back then, I recall ours was a mermaid. At times, to track our progress, I would borrow the road map, one of those immense pieces of fragile paper whose maddeningly intricate folds were designed by Indonesian monks as a form of penance for coveting an extra bowl of rice.

Traveling pressures like these could be exhausting, so I'd have to lie down across the backseat and, after we got the convertible, watch the clouds drift by above until the car's rhythm and the sun's warmth soothed me to sleep. Until I felt the car slow, when I instantly sat up, fearful of missing something interesting like a roadside Burma-Shave jingle or a new road sign. I remember coming off the George Washington Bridge onto U.S. 1 once and spotting a sign that read, SQUEEZE, which was eastern for MERGE.

In Boston, where I thought they drank an awful lot of tea for people who didn't look ill, the old highway's route has been changed over the years, according to the vagaries of construction and traffic flows. Now it is part of an expressway that ignores the historic road to Lexington, where Paul Revere perfected his messenger business; obviously, this was after the famous rider began making bells but before he got into life insurance and stainless-steel cooking pots. Modern 1 also runs beyond view of the *Constitution,* Old Ironsides, where on one family trip I walked the deck right where midshipmen my own age once took important orders and delivered vital messages that changed the course of a ship, if not history. Today, as part of an expressway, U.S. 1 flies by over working novelties like the Quincy Market or any old inn that might have survived succeeding governments' drive to renew.

In the 1920s, U.S. 1 skirted Boston, to wind along the scenic Charles River past several meat-packing plants, and the mind-packing factories of M.I.T. and Harvard University, through Cambridge, and along numerous local streets. It also cut through West Roxbury and jumped over Mother Brook, America's first canal. Dug in 1639, the waterway connected the Charles and Neponset rivers, making Boston technically an island.

But for a while now, the highway has followed the aging Southeast Expressway, I-93, south out of downtown past some vast natural-gas storage tanks before cutting back west to reconnect with the old stage route down below Dedham. There, it begins the brief drive south toward Rhode Island through countryside pocked with the familiar array of mini-malls and major malls that give way briefly around Wrentham to shady rural areas that can still seem like a winter wonderland after a thorough snowfall. In 1669, John Woodcock founded North Attleboro, opening a tavern that also served as one of several garrisons along the dangerous route from Boston to Providence. Woodcock frequently had fatal exchanges with local Indians, who wounded him seven times and then scalped his son, Nathaniel. Today, the inn in Attleboro is called the Pineapple and the large mall is named Emerald Square. The drive-in movie is dead.

The highway also passes through Foxboro, another of the once-separate, once-rural communities that have been drawn into a metropolitan swirl. Right alongside U.S. 1 in Foxboro is Sullivan Stadium, home of the New England Patriots, known as the Boston Patriots until team owners realized who was really packing the old stadium seats.

That stadium is a towering symbol of how big-city professional football teams have taken on a regional identity, thanks to highways and television, and followed their affluent fans out to the suburbs. The suburbs have the safe and affordable space necessary for sprawling commercial monuments like malls and stadiums, not to mention room for their endless traffic and accompanying tailgate parties.

The same suburban stadium scene is easily visible in New Jersey, where the Giants have dropped the words *New York* to play in the more modern, more lucrative Meadowlands complex, and on down 1 on the border between Fort Lauderdale and Miami, where the Dolphins built their own new suburban stadium near an Interstate exit. With a modesty typical of show biz, both Sullivan and Joe Robbie stadiums were named for team owners. In Baltimore, just off the highway, the Colts owner moved his football team out of the old downtown stadium to a new arena some six hundred miles west, in Indianapolis. That is a city that views cars as so important that they have their own museum, albeit a museum of hybrid racing vehicles unsuitable for any family vacations.

Whenever any family car gets tired or rusty or cranky and costly or simply unfashionable, the auto makers remain happy to make more. As one result, there are today more than 562 cars for every 1,000 breathing Americans. That is twice the number in crowded Japan, where no new car can be sold until the local police precinct has certified in writing that the buyer has an off-street parking space ready. Parts of U.S. 1 in Boston, the Bronx, Baltimore, and Washington could do with some kind of similar regulation. But they settle instead for simply having police tow away illegally parked cars after the fact, although authorities seem to experience more difficulty spotting abandoned vehicles. One of the major adjustments that 250 million Americans face in the crowded twenty-first century, it would seem after considerable wanderings along more than one overbuilt and underthought thoroughfare, is the growing scarcity of room for fresh starts, physical and psychological. With very few remaining physical frontiers to use up, we are going to have to make do—and probably redo—right where we are. All of which would seem to argue that so long as there is some kind of fuel to run them, the role of being a prime guardian of private space will continue to guarantee the private motor vehicle a prominent and revered place in a mobile American society.

Roger Williams Slept Here

RHODE ISLAND IS ONE of those Eastern Seaboard states that is so small, Rand McNally must print the state's abbreviation out in the Atlantic Ocean—RI. Good luck trying to ever get the mapmakers to use the state's official moniker: the State of Rhode Island and Providence Plantations. U.S. 1 meanders for nearly seventy miles across Rhode Island from Pawtucket in the heavily industrial northeast to the more rural southwest corner where the road enters Connecticut by a Mobil station.

It is not difficult to discern when U.S. 1 hits a city. First, the gas stations begin requiring payment in advance of pumping the gas. Second, virtually all of the trademark U.S. 1 shield signs disappear. In Rhode Island, the highway is better known as Broadway and much later as the Post Road. In most of the larger Eastern cities it visits, U.S. 1 becomes lost in such urban anonymity, like many of the cities' residents. To the millions who back their car out onto it, or park their car to wash it, or cruise the street to be seen on it, U.S. 1 is much more familiar as a local street and almost unanimously unknown as a major highway with a concrete heritage that runs to somewhere else. The major highway that has usurped 1's fame is Interstate 95, whose red, white, and blue shields are ubiquitous.

U.S. 1, on the other hand, is heavily used but poorly marked, plunging into a city on one side, virtually disappearing for a few miles through aging industrial areas, jerking one block this way and another block that way around the downtown, then splitting into two one-way streets near a bridge before merging again, unmarked, somewhere on the south side. There it widens slightly to wind through some suburbs and out into the countryside, regaining its broader shoulders and resuming its familiar highway identity.

"U.S. One?" said the Pawtucket gas-station attendant, who drove on it to reach the multilocked safety of his work cubicle, to sit surrounded by potato chips and bulletproof glass. "U.S. One? Gee, I don't know any U.S. One. Where is it you want to go?"

"U.S. One?" said the policeman patrolling downtown Providence. "To be honest with you, I don't really know where it is. Maybe it's Broad Street."

Roger Williams considered it good fortune and divine Providence that, first fleeing England and then Puritanism just up the road, he ended up buying some land from a couple of Narragansett Indian chiefs in 1636 in what became Providence. Roger Williams (not the pianist) held the philosophy of freedom of conscience, which did not fit into Massachusetts in those days. So Providence became the first American community to separate civil government from the church. This prompted one early Massachusetts writer to describe Rhode Island as "that filthy, nasty, dirty colony." But it also helped establish an important and distinctive tone for civic affairs on the American continent.

At first, Providence, some thirty miles up island-pocked Narragansett Bay, remained a backwater to Newport on the Atlantic. Three years younger, Newport, with its diverse immigrant population of Quakers, Jews, and Protestants, was Rhode Island's premier commercial and fishing center. However, after the Civil War, Providence's role grew as the political capital, while Newport became the capital of rich nonpoliticians who would hold a boat race or throw a party at the most expense for the least excuse.

Without the governor's really sincere welcome sign announcing the state's border, no one on Route 1 would know they were in a different state, until perhaps they paid the sales tax. The state is also famous for Rhode Island Reds, a special breed of chicken developed before serious farming became largely invisible in the Northeast. That chicken is now the state bird. Providence, where 1 jogs through downtown before crossing the Providence River to follow the coast more closely than I-95, now is home to the Paul Revere Insurance Group, the disability insurance arm of the billion-dollar conglomerate named Textron.

Pawtucket, which claims to be the cradle of the American Industrial Revolution, is home to the Pawtucket Red Sox, a Triple-A farm team of the Boston Red Sox. The city is also headquarters for Hasbro Industries, maker of G.I. Joe, Cabbage Patch dolls, Milton Bradley games, and Knickerbocker stuffed animals, which so successfully has tapped into the $225-a-year toy-buying budget of the average American family.

Pawtucket was an early home to water-powered cotton spinning, thanks to some industrial espionage work by Samuel Slater, a young English cotton-mill mechanic. In the late 1770s, it was severely forbidden to take any cotton-mill plans from England. So Slater memorized them and left to establish operations in Rhode Island. Slater also helped develop the concept of Sunday schools; thanks a lot, Sam. One of Slater's early financial backers was a man named Moses Brown, who along with Elihu Yale down the Post Road in Connecticut helped invent the Ivy League by giving money to a college. Yale started as a boy's school and ended up in New Haven, in 1701, while Brown began as Rhode Island College and stayed in Providence.

Previously, Pawtucket had been the home of the inventive Joseph Jenks, who patented the two-handed scythe and the first American fire engine, and made the dies for the first American coins, the Pine Tree Shilling. Jenks's successors saw a better future in defense contracts; they got into making ship anchors and muskets.

Although Rhode Island was the thirteenth state, it made its own declaration of independence from England two months before the official one. Close commercial observers could have seen that coming; British merchants could hardly sell their fine pottery to Rhode Islanders, who very much favored the inferior local pottery products. The coarse clay items from East Greenwich tended to fall apart, but "Buy American" was a popular slogan even then.

The Providence area was also the scene of the 1772 *Gaspee* affair, the other first violent incident of the Revolution. It seems that a British revenue schooner named the *Gaspee* had, under the guise of the commercially oppressive Navigation Acts, been harassing maritime businessmen south of town. In the tradition of isolated coastal residents, these local men did not regard ship hijacking and dodging import taxes as pirating or smuggling. When the *Gaspee* ran aground one June day, several boatloads of entrepreneurs, sixty-four good men stout and true led by another Brown (John), assaulted the ship, wounded the captain, and burned the craft to the waterline. This took care of not only the *Gaspee*'s troublesome enforcement activities around Warwick, it provided the opportunity for a modern tourist festival in June to balance against Newport's music and jazz festivals in July and August. Having seen, like Maine, the handwriting on the hull of the shipbuilding and other old manufacturing industries, Rhode Island's governor now prefers his state to be known as "America's First Resort." In a kind of silent agreement to disagree, the seaboard states only proclaim their own touristic virtues and have refrained, so far anyway, from pointing out the blemishes of their neighbors.

From Providence's downtown, an unmarked U.S. 1 wanders to the southwest along another one-time Paul Revere route through aging neighborhoods of gas stations, McDonald's, Bessie's Deli, and one-story groceries and stores whose hand-painted window signs in English and Spanish seek to attract customers from the most recent racial or ethnic group to populate the old streets. In southern Rhode Island, U.S. 1 has several local names. One of them is not Gilbert Stuart, the son of a snuff grinder, who went on to become a famous American portrait painter; he had to settle for having a side road named for him. But a portion of 1 is named Oliver Hazard Perry Highway, for the local lad who grew up to be such a quotable naval commander ("We have met the enemy, and they are ours") while facing the British in the Battle of Lake Erie in 1813, which came after Yale's founding,

if you're keeping track. With the proper battleground background on the evening news, Perry might have gone on to great things, or at least famous things, like the general he sent that message to, Benjamin Harrison. But, alas, the naval gentleman from Rhode Island would not be the first from his state to reside in the White House. He had to settle for being an inland hero.

Oliver Perry's little brother Matthew also took to the sea and made a name for himself far from his home state's rocky coast; Matthew Perry was the commodore who opened feudal Japan to the world in 1853, back when Americans thought they could teach Orientals a lesson or two.

Near Greenwich, Rhode Island (not to be confused with a different sort of Greenwich in the next state), the highway passes close to the gathering spot for a Tory mob (in America, Tories gathered in mobs, Patriots assembled in crowds) bent on erasing the subversive community of East Greenwich. It is a hoary New England tradition to keep track of where famous people slept instead of something they might have done while conscious. And so, although it was razed about one century ago, the William Arnold Tavern is known as the place where President Lincoln slept one night in 1860. Gone, too, are the numerous places where Ben Franklin is said to have slept, as well as Lafayette's room and Tom Paine's, and now we can all understand why there is no lingering political clan of Washingtons to revere and re-elect; George was never home to found one. But this Greenwich community itself has survived, its aged homes still lining the hillside up from the waterfront, while the old ma and pa clothing stores, with their faded brick fronts, struggle to survive in an era of discount stores. The elderly Kent Theater has now become the Kent I, Kent II, and Kent III Cinemas.

After East Greenwich, but before the University of Rhode Island campus, stands the Hill-Top Drive-In, overgrown and silent. No more will drivers ease their vehicles up the little gravel inclines so that their chromed grilles can pay obeisance to the movie screens towering above the countryside like commercial temples. No more will drivers position their left window by the metal stanchions holding the bulbous speakers with the malfunctioning volume buttons—and then forgetfully drive away, tearing out the speaker or possibly the entire window. No more will there be the inevitable latecomer who forgets to douse his headlights before illuminating the bottom half of the screen, igniting rolling waves of honking horns. No more will the refreshment stand dispense popcorn and soda pop in flimsy cardboard carriers to be precariously portaged back to the car, where couples could spend an enjoyable two to four hours, some of it watching the movie.

Drive-in movies are a business with a wonderful future behind it. Taking a family or a parent's car to the outdoor movies was once a very exciting novelty. Drive-in movies were invented in the 1930s, not far off U.S. 1 in Camden, New Jersey. At its peak in 1958, the country had 4,063 drive-ins, most of them in once-rural areas fast becoming suburbanized. The feature films, preceded by previews and possibly a newsreel, began at dusk on carefree summer evenings with couples and families, many of the children already in their pj's, staring through the windshield and possibly annoying clouds of whining mosquitoes. The appeal was an evening in the cooler countryside. The excitement of a new movie with scenes never before seen; the economy of buying one ticket per carload, the convenience of a baby-sitter-free night out; the privacy of a personal automobile in a darkened theater: They all combined to draw customers like moths to a white porch light.

But a new combination of social and commercial ingredients emerged in the last quarter of the twentieth century to overwhelmingly tip the balance of cinematic power away from the drive-ins. For one thing, the indoor theaters were all air-conditioned (and mosquito-free). The growth and variety of television and then cable television and then VCR's brought larger screens and theater-quality sound into private living rooms. Occasional gas shortages kept many people closer to home, where theater owners were building new multiplexes, which, though

ugly, were certainly sterile. They offered a wide choice of film fare every evening. Technological advances in projection and sound reproduction made much of the drive-ins' equipment obsolete. The housing developments and malls of the mushrooming metropolitan areas now completely surrounded most drive-ins, putting unbearable financial pressures on land values, especially in a business that, in the North anyway, could be open for business only part of the year. And the sexual revolution of recent decades eliminated the need to flee as far as the edge of town for carnal privacy.

As one result, by 1971 the number of drive-ins had fallen to 3,720 compared to 14,055 indoor theater screens. Nine years later, the drive-ins totaled 3,561; the indoors, 17,590. The United Theatre in Westerly, Rhode Island, down by the Connecticut line, was typical. Built in 1928 by Samuel Nardone, it packed in viewers from two states, especially on Friday and Saturday nights, when the streets in front of Westerly's downtown stores were lined with cars, while the lanes in both directions were filled with through traffic. Sometimes the United's screen was put away in favor of live acts making the theater circuit.

But the zoning laws changed, making it cheaper to build new stores in the developing commercial strips on the edge of town than it was to restore the old facilities. Updated fire codes also made it necessary, but too expensive, to renovate the old buildings downtown. The United had one owner from 1928 until 1981. It has had several since then. Today, the theater stands vacant—and available—while a mile away, the Westerly Cinema's three screens pack in those customers who haven't rented their own movie for the night for less than the price of a single ticket.

By the last decade of the century, the population of indoor theater screens like the Westerly had soared to 23,132, while barely 1,000 drive-ins survived, many of them in the South. The other outdoor screens were torn down for new developments or left to stand like gargantuan windscreens for the weekend flea markets that sprouted on the gently inclined rows of gravel below.

George Washington Grew Indignant Here

COMMUNITIES ON THE SCENIC Connecticut coast developed individually to serve the economic and social needs of their adjacent agricultural backcountry areas in a kind of vertical inland society. There being little more than uncertain paths between these communities, they had a strong sense of isolation and independence, and, some say, downright orneriness. The Indians thought they owned the area. Then the Dutch thought they had claimed the lands as early as 1614. Then along came those Massachusetts Puritans again and straightened everybody out, one way or another. To this day, despite its location next door to New York, Connecticut still seems to regard itself as more New England, from its quaint town meetings to its television sports coverage, which assumes an allegiance to Boston teams (except for hockey in Hartford).

Settling such a wild place required a firm hand and an unbounded ego. According to the records of one very democratic meeting in early Colonial times, a Connecticut church settled any doubts. "Resolved," the minutes recall, "That the Earth is the Lord's and the fullness thereof; Voted, that the Earth is given to Saints; Voted, that We are the Saints."

A few deals were done with the Indians, generally involving a pile of blankets and some trinkets in exchange for all the land within a day's walk or some such helpful measurement. If there was any disagreement over price or ownership, the matter was settled like Christian gentlemen: The tribe was annihilated. The Dutch and English, who needed the community of Cos Cob as a future home for New York City railroad commuters who would pay extra for that name but would not be told that it originally belonged to some elder savage, got together in 1644 to take care of the local Siwanoy tribe. Not a man, woman, or child survived, which helps explain one contemporary account's observation: "Nor was any outcry whatsoever heard." Back up the coast, the town of Mystic erected the Mason Monument in honor of Captain Mason and that glorious day when he and a few dozen settlers burned six hundred Indians while they slept. To be sure, a few escaped, but Mason got them later at the Great Swamp Fight near Fairfield. Praise the Lord!

By the Revolution, the lower Boston Post Road (as opposed to the upper one that cut up from New Haven to Hartford to Worcester and Boston) was rough but adequate enough to let ten congregational ministers, disturbed at the ridiculously liberal teachings being dished out up at Harvard, move their combined libraries from Old Saybrook to New Haven for the 1701 founding of the "Colligiate School," later known as Yale. Paul Revere made good time on the same road to Philadelphia carrying word of the Boston Tea Party. A not-yet-traitorous Benedict Arnold and his boys quickly marched up the road when news of the Lexington skirmish reached the Arnold family pharmacy in New Haven. The Boston Adams clan traveled back and forth on their journeys of ambition. Ben Franklin and his doughty daughter covered many miles on this road with the postmaster installing stone mile markers, the point being to settle once and for all nagging disputes over postage, which in those days was paid by the mile. B. Arnold used the same roadway to lead British troops against several shoreline communities. Stagecoaches carrying fancy written names like Regulator and The Hero jounced their passengers over the ruts, much as do today's semitrailer trucks with their own names painted on the windscreen.

And G. Washington, who preferred to ride his own horse, made several journeys along what would become U.S. 1. Of course, many businessmen and society ladies clamored to provide food and lodging for the nation's chief executive. This was before the country's leaders needed ethics laws to keep them out of trouble, so Washington decided on his own that it would not look proper for a president to choose some citizens to stay with, while rejecting the offers of other voters. He was also concerned not to appear the least bit regal in demanding accommodations. So, being on an expense account anyway, he dined and slept here and there at public inns. However, during one overnight stay at Clark's Tavern in Milford, the commander in chief, perhaps tired from a full day on Connecticut's highway, grew indignant when he was served a dinner of bread soaked in milk. It seems the waitress brought him a pewter spoon, which would not do for his lips. The president gave the waitress a shilling to borrow a silver one from the nearest minister.

Much later, the road was used to haul granite from Milford's famed quarry up to adorn Boston's Public Library and down to New York City's railroad stations and Washington's Corcoran Gallery. David Bushnell and Sergeant Ezra Lee traveled along the old highway in 1776 to launch the *Turtle* at Old Saybrook. Designed by Bushnell and piloted by Lee, that ship was the United States's first submersible. And it launched its first attack that same year in New York harbor, delivering a bomb to the bottom of H.M.S. *Eagle,* which didn't sink. To this day, that same section of Connecticut coast remains home to the American Submarine Service. Groton is headquarters for the Atlantic submarine fleet, as well as for General Dynamics' Electric Boat Division, which launched the first diesel-powered sub there in 1912, and the first nuclear-powered sub, the *Nautilus,* in 1954. Across the Thames River stands the Coast Guard Academy; and the bridge, which U.S. 1 and I-95 share, can sometimes provide a view of those sleek cigar-shaped ships heading out to sea, as the whaling ships did before.

There is a brief sylvan section of U.S. 1 near Mystic, where motorists can glance to the side and see the graceful masts of a whaler, a square-rigger, and a fishing schooner safely moored now at Mystic Seaport, awaiting the next tide of tourists. But from Groton-New London on down

well into Virginia, U.S. 1 has also become home to perhaps more strip malls and car lots per linear foot than any other American highway. They begin with their own exit (New London Frontage Roads Shopping Malls) and the Show of Hands Nail Salon and move through an eye-glazing array of modern roadside Americana—Burger Kings, McDonald's, Mobils, Dunkin' Donuts, Radio Shacks, Pizza Huts, and Burlington Coat Factory Warehouses, all punctuated by aged cemeteries whose caretakers must be off this year. Right about here, the Italian restaurants take a momentary lead in the U.S. 1 ethnic dining sweepstakes, to be thoroughly pummeled later, of course, by the Mexicans.

A few of the older establishments have survived. Bishop's Orchard in Guilford is crowded by gas stations but still produces tasty fruits. In Madison (FIRST SETTLED 1650), the Wayside Motel and diner are gone, but the sign remains: OLD-FASHIONED DOUBLE SCOOP ICE CREAM. The Nautilus Fitness Center and racquetball court are prospering, as is the Carisma Car Wash in Saybrook, at least after a storm. But Elaine's Hair Salon in Westbrook must also sell Iams pet food to bring in more customers and Marty's Hot Dogs has colored models of all his condiments up on the roof, which is a big help.

Much of the adjacent plant life has been removed or has perished in the steady flow of fumes that U.S. 1 produces in congested urban areas where traffic is stop-and-start and trees get in the way of signs. Eight hundred miles out of Fort Kent and its surrounding forests, a lone roadside pine survives in East Haven, many of its branches and all of its top missing. Those limbs that remain are curling up.

Some of the communities have worked hard to maintain their small-town identity. And they remain as nearly identical beads in a necklace of civilization along the shoreline. They are still centered on a rectangular village green with white churches around the edges. Few law-abiding people lounge around on the greens anymore and no one grazes their cattle there. Nonetheless, many of the surrounding buildings' small Colonial windows survive; the panes were made small then not because it was the fashion but because that's as large as anyone could make them.

After the Indian business was settled, Fairfield (so named because Roger Ludlow, another one of those Massachusetts arrivals in the 1630s, thought the surrounding salt marshes were fair fields for grazing) had problems with witches (also with Aaron Burr flirting with John Hancock's fiancée, but that's another story). Goody Knapp was believed to be prominently evil in Fairfield and got hung. But contrary to modern belief, it was her sisters in darkness, Mercy Disbrow and Elizabeth Clawson, and not Joseph Heller, who invented catch-22. It happened in a pond now gone from the western end of Fairfield's green. There, suspected witches were tied up and tossed into the water. If they sank, they were obviously innocent, though soon dead. On the other hand, if they floated, they were guilty and soon dead. According to one account, Mercy and Elizabeth "buoyed up like a cork."

But appearance is not the only thing that has changed along the old road. The industrial foundation that provided so many generations of income and emotional stability to communities such as Pawcatuck is being replaced by a service economy that does not pay as well, in money or personal satisfaction. For instance, in Pawcatuck, by the Rhode Island border, the Harris Graphics Corporation closed recently after 134 years of manufacturing printing presses. Generations and entire families worked there and bought their groceries and clothes and maybe walked home for lunch or, more recently, ate over at Elva Devaney's Whistle Stop restaurant. They got good money to feed their families and maybe a little extra to buy one of the numerous boats that line the Connecticut shoreline under blue tarpaulins.

The problems seem rooted in the kind of revealing corporate machinations and local pain that have become characteristic in the last decade. In 1980, the managers and investors of the Harris Corporation arranged for a leveraged buyout of their printing-equipment division, borrowing the money to buy their own company by using the firm's assets as collateral and its own income to make repayments. This created Harris Graphics, also a financial mire. In 1986, Harris Graphics was sold to a Chicago firm, with the sellers realizing immense profits. The sellers included Ivan Boesky, later convicted of insider trading, and Michael Milkin, who perfected the junk bond as a tool of financial rape and was convicted of mail fraud and conspiracy, among other dealings. (He got a ten-year sentence and a fine of $600 million [!], which still left him as one of America's richest men.) Two years after that sale, the Chicago owner moved some of Pawcatuck's key departments to a New Hampshire plant and sold the money-losing company to a West German firm. Less than two years later on an autumn Wednesday, all seemed well when Neil Mackenzie, Jr., a third-generation employee, and the rest of the second shift went to work. Three hours later, they were told to assemble their toolboxes and leave. The plant was closed. All the jobs were gone for good. The German owners provided generous severance pay and retraining, and helped arrange new job interviews. But for the nearly one thousand former employees, and much of the town, their confidence in the future and themselves was shaken. And they had an inkling of a different, depersonalized economy developing along the roadway with emphasis not on production but profitable financial maneuverings.

New Haven, of course, is not new to change; Charles Dickens would hardly recognize the city's Hillhouse Avenue, which he called the "most beautiful street in America," with no record of his tongue having been in his cheek at the time. In 1662, Charles II had granted a new charter to Connecticut, which had been founded twenty-four years earlier by some Boston refugees. The new charter allowed the Colony of Connecticut to absorb the Colony of New Haven, which had been briefly known as Quinnipiack. As a sop to its lost independence, New Haven was made cocapital with Hartford for a while. Inland Hartford's future would be as a state political capital and national capital of insurance.

New Haven's future, along the coast and the coastal highway, was as a port and industrial center and a center of invention. Not only would New Haven become the first planned city in America but its residents would develop many things, from match-dipping machines to corsets and football shoes, from the first electric elevator and the automobile self-starter to the accidental invention of rubber vulcanization when Charles Goodyear spilled sulphur and rubber on a hot stove. For a while, James Jarvis of New Haven ran the nation's unofficial mint, having received a contract to make three hundred tons of coins, including the Franklin cent that Ben designed. (On one side were thirteen joined circles, while the other showed a sundial, the word *Fugio* ["I fly"], the year 1787, and the motto "Mind Your Own Business.")

New Haven was also home to famous clockmakers, the publishers of Webster's first dictionary, and to Dr. Joseph H. Smith, the first reputed to have used nitrous oxide in dentistry (in the single month of June 1863, thanks to laughing gas, the New Haven Convention and Visitors Bureau claims that he extracted 1,785 teeth. That's one every twenty-four minutes around the clock). New Haven's Yale also became the first recorded victim of a no-hitter when Princeton's Joseph McElroy had his best day on the mound, May 29, 1875.

Perhaps New Haven's single-greatest industrial contribution, however, was by another Massachusetts native, Eli Whitney. He is most famous for inventing the cotton gin. That machine could clean cotton as fast as fifty men. It revolutionized southern agriculture (and American exports) all the way down the nation's main highway and would help make the slaves obsolescent. But patent infringements and lengthy legal fights, even in the late eighteenth century, cost Whitney any financial success. So, hard by New Haven, Whitney built the nation's first company town. And there, in his worried haste to meet a government contract to produce ten thousand muskets in the incredibly short time of two years, he invented the concept of interchangeable parts, revolutionized manufacturing, and, not incidentally, set the stage for Henry Ford's assembly line. This helped create Connecticut's continuing role as the nation's arms maker, a position solidified by Samuel Colt and his perfection of the re-

volver, which gained instant fame and sales for its killing power in several successful wars early in the nineteenth century.

New Haven and its prominent highway also played a little-known role in the development of a vital resource, for better or worse. It was a chance meeting downtown in the old Tontine Hotel of James Townsend, a prominent banker, and Edwin L. Drake, a self-appointed colonel and person of immeasurable determination, that led in 1859 to the drilling in western Pennsylvania of the first successful oil well. Once this New Haven pair had uncovered ample reserves of that volatile gooey substance, it was left to two ruthless Clevelanders, John D. Rockefeller and his little-known partner, Henry Flagler, to find a profitable use for it. The impact of Rockefeller's Standard Oil trust and his resulting family fortune are duly famous. Less well known is that Mr. Flagler's parallel fortune and personal vision enabled him to become the first man since Ponce de León to see Florida's swamps as a destination for anything other than mosquitoes, alligators, and condo-eating hurricanes.

The United States had bought Florida from Spain in 1819. With the loss of 1,500 soldiers and the help of Samuel Colt's revolvers, Andy Jackson and other military leaders had pacified that swampy isthmus so it could become a theme park and spring-training site someday. And there was also a certain old coastal pathway that could help deliver Northerners to this tropical would-be paradise. All sleepy Florida needed was a dose of twentieth-century marketing skills—Flagler communities, extravagant Flagler hotels, a Flagler railroad connecting scenic tropical islands with an endless series of bridges running more than one hundred miles out into the Gulf of Mexico, perhaps a major highway running down Florida's coast, and some roadside orange-juice stands, which Flagler did not own but didn't mind. Ultimately, Flagler's Florida vision and all those domestic immigrants to the Sunbelt would drastically change the face of American society and politics, not to mention winter vacations and retirement.

What these southbound and suburban-bound immigrants left behind were bands of aging cities like New Haven, whose cheaper housing and fading factories seemed to offer a better future for many fleeing the fading farmlands. But streets paved with gold were as hard to find in urban America as pirate treasure on the beach. New Haven, like many other cities, beginning in the 1950s, went through a well-intentioned era of urban renewal bulldozing, which may have made the downtowns look better but didn't solve the social problems, as the epidemics of drug abuse revealed a generation later. Today, old industrial New Haven's largest employer is not a factory; it is Yale University. And while old U.S. 1 has lost its identity—and its own downtown bridge to rust and rot—it still struggles through the city known, in some neighborhoods, as Forbes Street.

About the time that the bulldozers were growling all around at full throttle, the old Church of the Epiphany at the corner of Forbes and Stiles streets went out of the religion business. Its patrons were either dead or in the suburbs. Their replacements had their own storefront places of worship. So the old church on U.S. 1 was sold. The new owners had work crews take out the altar and the pews. And they had them knock a large door in the back wall so that the Eastern Plumbing and Heating Company could park its fleet of trucks inside the old sanctuary, free from roaming vandals. "This way," said Paul Callahan, a plumber from Miami who wanders the country from construction job to construction job, "we figure at least our trucks won't get hit by lightning."

Before the railroad, New Haven was often the jumping-off point for New York–bound stage passengers. New Haven was among the busiest ports in New England (it still is, especially given that Connecticut is one of the few seaboard states without its own oil refinery). Levi Pease was probably the first organized stagecoach operator on the route. At New Haven, some of his riders would transfer to a sloop for the sail down Long Island Sound to Gotham. This was quicker and much safer than the tortuous trek by land over the rivers down the cliffs and through the shore's swamps, which were so secluded that locals could successfully hide from British troops two of the English judges who had condemned Charles I. The Connecticut coast was also infested with highwaymen, especially around Norwalk, which today specializes in car dealerships. Norwalk's claim to fame (not counting Stew Leonard) was being the jumping-off point for the ill-fated 1776 spying foray over to British-held Long Island by a new Yale graduate. The Dutch schoolteacher disguise didn't work. But no one in Norwalk believes that Nathan Hale's willingness to give one life for his country had anything to do with avoiding a return to Norwalk.

A ferry from the Connecticut shore over to Long Island still operates from Bridgeport, which is the state's largest city and hometown to a number of people who went out to influence American life, people like Walt Kelly, the wry cartoonist who created little Pogo, who lived way, way down U.S. 1. Every weekday about 100,000 people ride the twin pair of railroad tracks along the Connecticut coast through this affluent southwestern corner of Fairfield County, one of the country's richest and one of the most concentrated corridors of WASPs in the nation. Stamford, which had only thirty residents as late as 1641, grew to more than 100,000 with many originally coming as commuters, though the city is more independent now. The commuters, executives and executive wannabes, go all the way into Manhattan and then back out again at night, some traveling more than fifty miles each way, although the Bar Car is open at night, thank goodness. Along this section of coast, the tracks, the interstate, and the lanes of U.S. 1 twist back and forth over each other past Pizza City and Bed City and through towns like Stratford, Southport, Westport, Darien, Rowayton (the late Indian chief), and Stamford.

Then there is yet another Greenwich, a fifty-one square-mile enclave of affluence where about 60,000 people try to get along with only 52,000 cars, 242 snowmobiles, 13 private schools, 181 horses, and a median household after-tax income creeping up on $60,000.

Greenwich used to be called Horse Neck until town fathers decided that would not do for a community where so many residents felt the need to hire help to answer doorbells.

Greenwich's newcomers like to think of themselves as old money. Back when New York and Connecticut couldn't agree on where their border was, the community had real image problems; it was known for unchecked drunkenness and as a popular destination for elopers and runaways, both slaves and children. Nowadays, eighteen-year-olds in rented formal wear from surrounding towns run off to Greenwich's Hyatt for their senior proms, while proud parents take snapshots and smile in the ignorance that survives until fall's first freshman tuition bill.

Today's thick local traffic and the time it takes to creep through town after town on U.S. 1 might suggest that Connecticut is much larger than the nation's third-smallest state (the smallest, of course, is Rhode Island, which is also on U.S. 1, and the second-smallest is Delaware, which ought to be on U.S. 1). But in the tradition of modern American suburbia, much of the traffic has little to do with going into the nearest huge metropolis; many are commuting to and shopping in the economically independent municipalities, which contain so many headquarters of well-known corporate giants like General Electric, Champion, GTE, Xerox, American Brands, Continental Can, General Signal, and Pitney-Bowes. Indeed, Fairfield County is home to more Fortune 500 companies than anywhere outside Chicago and New York City. Ten thousand Greenwich residents commute out of town each workday, many into New York. But twice as many workers commute *into* Greenwich daily.

On the weekends, a fair number of drivers on U.S. 1 are visitors, which the grocery stores can ignore and the quaint craft shops and theaters adore. Some towns like Westport (average house price—$492,000) have become home to a well-heeled share of celebrities who want to be near the Big Apple but outside the urban orchard. They include Paul Newman, Joanne Woodward, and Rodney Dangerfield. David Letterman, who lives nearby, suggested a possible new motto for his state's license plates: "Connecticut: The Second C Is Silent."

Dominic Brigante has another suggestion. He lives in Darien, which is pronounced Dairy Ann, and don't you forget it if you're going to pay those kind of property taxes. Darien is where the patriots (the soldiers, not the football team) launched whaleboat raids against Long Island loyalists. But now it is home to Dominic Brigante's gas station on U.S. 1 in a two-hundred-year-old shed that began life as a carriage shop, moved into trucks, and today dispenses gas and tire repairs to cars, many of them very expensive. The station is so old that Dominic found three roofs when he tried to fix one leak. And the car lift is outdoors because good weather was the only time people went driving in the years when the lift was installed. Anyway, Dominic is from Italy. He had another gas station, down 1 in Port Chester, New York, but he sold it to move to Darien—kind of moving upscale. And now he's sorry. Port Chester residents take care of their own, he says.

"Darien is lousy," he adds. "Lousy people. They are rude to themselves. A bunch of WASPs. I have nothing against them. But they should shop Darien. Darien is a bunch of ten-cent millionaires. I live there. I know what's going on. Darien customers will drive around until they run out of gas, trying to save two cents a gallon. These rich people go to Stew Leonard's to save two cents on milk."

Stew Leonard's. Now, there's a Route 1 institution. It is a one-of-a-kind grocery store. And he is a former milkman, one of those sly, now-rich people who can turn the adversity of a highway displacing his dairy farm into the prosperity of opening a dairy store to milk the passing traffic. He breaks all the traditional rules of food merchandising. And people love it. They come to his Norwalk store from all over, even two states away, knowing full well they won't find much variety. What they *will* find are milk and meat and vegetables, fresh, and probably a fair bit cheaper than anywhere else. And they'll probably have a chuckle or two, as well. And a free one-dollar ice cream cone, if they buy enough groceries. But what family of shoppers is going to voluntarily live through the ordeal of just one kid getting an ice cream cone? Right! None. So by giving away one cone, Stew sells several more. Got the picture?

In many societies, shopping for food is a necessity. For some reason, in the United States, shopping is also an outing, an event, an excuse for couples and families to do something benign together. "I try to make it fun," says Mr. Leonard. So animated milk cartons talk to customers. Fresh young employees seem to pack the aisles, helping everyone, stocking shelves, and dressing up as cows or chickens. To keep in touch with at least some of each week's 100,000 customers, Mr. Leonard may even don a chicken outfit to stroll the aisles, where, instead of the usual 25,000 items found in a usual supermarket, Leonard's has several hundred, and even fewer brands. But they're cheaper. Outside, on the way to the satellite parking lot (Stew's is so busy, he needs his own set of stoplights on U.S. 1), the little people can feed the animals at a small zoo with the feed on sale inside. Stew borrows the ducks, chickens, and goats from area farmers for a few weeks on the provision that his customers will help feed them.

Stew also does something revolutionary in modern merchandising: He listens to his customers. No, really. See, he has this suggestion box, which he actually reads. One shopper suggested customers be allowed to pick their own individual strawberries instead of taking a pre-wrapped box. Stew's helpers said customers would eat a lot of the loose berries and not pay. They were right. But the ones who did pay for the fruit bought far more than they had been. More profit.

Stew had been selling turkey dinners, three turkeys' worth a day, with vegetables, wrapped and refrigerated. One lady said, "Sell them at the hot-food bar." Stew tried it. Demand rose to twenty-one birds a day. Others complained about paying $2.99 a pound for meals with high-calorie gravy; others said there wasn't enough gravy. Stew put the portions on the side. Sales rose to fifty turkeys a day. In fact, in a business where total sales might run up to $500 per square foot of store space, Stew Leonard's does $3,100.

This Way to Egress

THERE IS NO EVIDENCE whatsoever that the successful Stew Leonard is related in any family way to another famous and very successful Connecticut businessman and entertainer, one Phineas T. Barnum, by name. Yes, Barnum of Barnum & Bailey fame, the Greatest Show on Earth. The man who dubbed his six-ton elephant Jumbo and thereby gave the English language a new word for very large. The very same man who once met Charles Stratton, a dwarf from nearby Stratford, Connecticut, and dubbed him Tom Thumb. Barnum was a native of Bridgeport (so was Jumbo during the winter break). In fact,

Barnum was Bridgeport's mayor for two years, and a state legislator, and a philanthropist, and a lot of fun.

There is no evidence that Barnum ever said "There's a sucker born every minute," a quote attributed to him by a former employee long after the showman's death. But Barnum did say "The bigger the humbug, the better the people will like it," a maxim that advertising and television and several other American industries have enshrined in their corporate hearts. Quite simply, Barnum was, as Glenn Collins has written, "the Michelangelo of buncombe, hokum, hoopla, and ballyhoo." He perfected the freak show, popularized the modern public museum and public concert, designed the most imaginative and misleading signs, and turned the circus from a polite collection of understated acts into the most entertaining extravaganza of felicitous festivities, awesome arrays of costumed finery, and death-defying deeds of derring-do, arranged before your very eyes in three (not one, not two, but three!) raucous rings of merriment, mirth, and miracles that had ever been seen before by sputtering circus purists and millions of delighted ticket buyers.

And now, ladies and gentlemen, let me direct your attention to Barnum, perhaps the first Teflon-coated man in the entire history of the world. If his famous "Wooly Horse" turned out to be a regular horse with cotton glued on, people just shook their heads without any real Ralph Nader outrage. If his famous "FeeJee Mermaid" turned out to be the top half of a monkey sewn on to the back of a fish, well, where's the harm (in the days before animal rights, anyway)? And if Joice Heth, the genuine 161-year-old slave nurse of General George Washington, the Father of Our Country, was barely half that age when she died, who forced anyone to buy a ticket?

People got their money's worth anyway viewing Anna Swan, the Nova Scotia giantess, and Madame Josephine Fortune Clofullia, the bearded lady, not to mention Chang and Eng, the twins from Thailand who shared a liver and talked about what it was like to live joined to another human. They popularized the term *Siamese twins*. (But they never did talk about what it was like both to be married to a different woman). Certainly, no one forced the newspapers to write about all this, although Mr. Barnum was never against publicity, any publicity. As P.T. himself once put it, "Every crowd has a silver lining."

Barnum was also a newspaper editor, a brief convict (for suggesting that a local minister had practiced usury), a lecturer, and a real estate tycoon, being what Robert S. Pelton, curator of Bridgeport's comprehensive Barnum Museum calls "the Donald Trump of Bridgeport."

In that tradition, Barnum made a fortune, lost a fortune, and made it back. He brought Jenny Lind, the Swedish Nightingale, over from the old country and up the highway from New York. Out of his simple desire to unite two people in love in 1863 and to share his joy with two thousand of his closest friends, Barnum arranged for the ceremony to be at a large New York City church, where the throng watched the diminutive Mercy Lavinia Warren Bump become Mrs. Thumb. There was a little something in the newspapers of the following days to break up the gloomy war news. And, as it happened, the wedding of two dwarfs didn't hurt ticket sales when Barnum's show hit the highway that spring—which is how it went in pretelevision show biz, even if your name wasn't Geraldo.

As a fitting tribute to a native son who did well, Bridgeport to this day has a Barnum Tile Store over on Barnum Avenue and a Tom Thumb Florist up on North, which is now U.S. 1 and where Barnum and Thumb are buried. Also on North is the Connecticut Community Correctional Center, which they started building nine years before Barnum died. They're still not done. Once, authorities performed executions on the Post Road, leaving the victims to hang for months as decaying symbols of the wages of sin. Over the years, the Northeast, as opposed to the other end of U.S. 1, has gotten away from the death penalty. But thanks to drugs and guns and money, the prisons have had to grow, of course. Bridgeport's still is.

The tall fence and razor wire now surround a large city block. Every year about ten thousand men come and go there, as if part of a ritual, some attending trials, some spending long terms as residents of North Avenue. They can look out the bars and see a few trusted prisoners mowing the lawn while today's visitors, always wives or women, anyway, with little babies, line up to be checked for metal. They can't see the Balooney Tunes helium balloon shop from there or the post office in the strip mall or Mike Bogoslawsky, the Chevy dealer who says,

"I'm in your corner." But they can see the billboards advertising the private alcohol-and-drug-treatment centers springing up all over.

One of the prison's occupants who can leave at night is Michael Chernovetz, the son of a plumber, who decided that pipes were not his calling. He followed a friend's suggestion to look into prisons, which wasn't as much of a growth industry back in the sixties. Steady work. State job. Pretty soon—well, it seems soon looking back, although it wasn't living it day to day—Mike Chernovetz, the corrections officer (that's union for prison guard), was Michael Chernovetz, the warden. Now after his morning commute down U.S. 1, he watches new generations of young men enter and reenter his facility week after week. He's seen them pass through at various stages of their criminal career and they've seen him move up from corrections officer to lieutenant to captain and now warden. Many of the inmates are unaware of who their fathers are, let alone how exactly their misdeeds were wrong enough to win some time behind the razor wire.

"Today's prisoners are one tough breed of guys," the warden says. "They live by a street code that says, 'You do me or I do you.' It's very macho and very dumb. Prisoners used to be more upstanding: 'Yeah, I did the crime and you caught me.' Now everybody's in the fast lane. They have no sense of what they're doing to themselves, to society, to their victims. They give not one second's thought to blowing someone away—you, me, a cop, anyone. No thoughts beyond money and this afternoon. The young ones are the worst. They've never had anybody care enough to teach them any discipline."

Felipe B. was twenty-five years old and five months into a three-year sentence for possession of drugs when the warden ran into him that morning not far from his cell on U.S. 1. Like his brother, Felipe had smoked marijuana from age fourteen. He did cocaine daily for years.

"But I quit last year."

Did that have anything to do with his entering the Bridgeport facility?

"Yeah, I guess."

Had he ever had a job?

"Yeah. I sold Street Justice. It's a brand of coke. I'd make two, maybe three grand a week on seven-thousand-dollar sales. Sure, sometimes people take a shot at you. But when you're dealing, you're always packing, too, so you can send one right back, you know?"

No, I said, I meant, had he ever had a legal job?

"Oh. Yeah. Well, I was a housepainter for a while and I washed dishes part-time."

How much did he make there?

"Painting, I got ten-fifty an hour."

One day, after two consecutive urine samples tipped Felipe's parole officer to his continued drug use, Felipe just stopped reporting in. He continued to deal drugs. The parole people didn't catch him for five months—overworked staff.

"Dealing is dangerous," said Felipe, who has six children he knows about. "But I liked the challenge. I had good suits, a lot of gold, a brand new Toyota. I put five thousand dollars in stereo equipment in it. Took out the backseat. Sixteen speakers. I tell you, it felt real good to throw thousands of dollars down in front of that car dealer dude."

It didn't last long. The car was seized three months later and sold. Felipe looks over at the warden. "I know now it's not worth the money," said the prisoner. "It's better to earn two hundred dollars a week honest than two thousand dollars dirty. If I'd a-worked for that car, I'd still have it."

Not all prisoners spend as long inside. American society has yet to develop a comprehensive criminal-justice punishment system, as opposed to just more prisons. We're building more than $2 billion a year in new prisons and shipping off more convicts than the facilities can handle because the voters are afraid of crime and the politicians and judges are afraid of being unelected. So they order up more cells and stiffer sentences because that's pretty quick, as government actions go, and any serious solution would take longer than one term of office and might be misunderstood as coddling.

While many agencies are involved with sending along more prisoners, no one agency or person in the United States is accountable for seeing that prison works, or possibly developing a range of punishment alternatives, like parents have, that would provide some certain, swift, though not necessarily harsh punishment. The idea being to teach deterrence, not cynicism.

Today in actuality, there is prison or there is nothing, a lobotomy or an aspirin. Now, with court orders governing prison overcrowding in virtually every state, even getting a prison term doesn't guarantee much punishment. Many convicts are set free after serving only a fraction of their sentence. A majority are chased again

for something else. They're caught. They're investigated. They're tried. Taxes pay for all this. The next time, to show them that society really means business, they're sent back to prison with an even longer sentence in their file. Building more cells hasn't stopped crime, so maybe building more cells will.

The warden describes a typical case. "We nab a kid for selling dope. He should do a few months' time. But we're real crowded. He's not violent. It's his first or second offense. So he's out early. We had no money to train him. All we did was process him, feed him well, treat him medically at taxpayers' expense, and protect him from his enemies on the street for a few weeks. Can you blame him for thinking, Hey, if they're not taking this seriously, why should I?"

Duncan Hines Ate Here

THERE WERE FOUR PEOPLE in my family—Mom, Dad, me, and Duncan Hines, who never met any of us.

Duncan Hines had been everywhere, like Howard Johnson. He wrote books about restaurants and, later, motels and, even later, a guide to carving, back before knives had cords. Duncan Hines did not provide denunciations; if a place wasn't good, it didn't get mentioned, period. It was a powerful omission, like not being invited to a party in a small town; you got the message.

Mr. Hines did not write windy reviews with turns of phrase designed to stun the reader. Wines weren't nutty or fruity to Duncan Hines: no *sauvignons* with noses. Duncan Hines thought down-home was best, plain and simple like a decent berry pie. For a Kentucky boy, he sure seemed to like seafood. He most loved New England food and was impressed with family-run places that had a personal stake in quality. He would have loved Moody's

up in Maine, which may be why I still stop at those kinds of places.

"Recommended by Duncan Hines" was a powerful endorsement in my family, better even than a president's endorsement, since he might be in the wrong party. If Duncan Hines said it was good, it was good—like the people down the road back in Ohio that were so proud of their good jellies and jams that, against all the experts' advice, they put their family name on the label, Smucker's. My mother bought Smucker's before they got famous. And she followed Duncan Hines's recommendations religiously because he had obviously been everywhere, and if he hadn't been in some town, then it wasn't worth our lingering there, either.

In those days, it was written on a granite tablet somewhere that all members of all families ate together at the same time every day, unless one of them had died that afternoon, and then he was excused. I can't prove it, but I believe that the government set all its clocks then according to my mother's mealtimes. I picture a federal functionary on our phone in the kitchen one evening a month, his right hand raised, his eyes watching my mother intently as she prepared dinner. As the moment neared, he alerted the astronomers on the other end. Dad and I were at the table, as usual. "Ready?" the man would say into the phone as Mom turned from the counter. He'd watch her very closely as she brought the plates nearer and nearer to the table. Tensions would rise. All second hands were poised at twelve. She was almost there now. "It . . . is," he would say, ". . . six . . . thirty . . . right . . . NOW!" His arm would drop. All federal clocks would start. And we would eat.

On our countless car trips over the years, about an hour before mealtime, Mom would reach toward the glove compartment and say, "Let's see what Duncan Hines has to say today." The first couple times, I thought she had this guy in there and he was going to jump out, like a liberated genie, and say lots of profound things. But it was just that Mom kept her copy of Duncan Hines's *Adventures in Good Eating* in the glove compartment, so-called because in the early days of motoring my uncle Jim stored his gray leather driving gloves there. Mom would shuffle through the shiny paper pages to the appropriate region. And Duncan Hines would determine where the Malcolm family ate—which was very important by 6:25.

I had the impression that this man, who did nothing but eat and write, might have rushed his latest recommendations to Mom's side of the car the night before while we were sleeping at the motor court. Or if the dinner hour had somehow sneaked up on us, Dad would pull our bulbous Plymouth into some restaurant parking lot, our big balloon tires crunching the ubiquitous gravel beneath. And Mom would check out that particular place with Mr. Hines. As a result, Duncan Hines had been to every place my family ate. We never actually saw him, mind you. But you could tell he had been there. We'd

walk up to the door and there was his name: "Recommended by Duncan Hines." Must have just missed him. Where had he sat? My plate might have been his? Even the spoon.

One time, I thought we had come particularly close to Duncan Hines. I asked the restaurant owner, "Is he still here?"

"Hello, little boy," he said, smiling over at my parents. "Is who still here?"

"Duncan Hines," I replied, all chipper. "Is Duncan Hines still here?"

The man blanched.

"Duncan Hines is here?" he shouted. "Where? What?" His head began to scan the restaurant so quickly, I thought it was whirling. My father gave me one of those "Now look what you've done again" looks. By now, a fair number of other diners were actively looking around for Duncan Hines, too. But none of us knew exactly what he looked like. So he got away again.

Mr. Hines's green paperback book was expensive— $1.50—and was arranged by states, then by towns. It had each establishment's address, brief directions, its hours or off-season (many closed after October 1, you see) and four-digit phone number, what the place was most famed for, maybe how unique the furnishings were, and the price ranges for each meal served. He might mention a particularly astounding menu item, but I checked and he never reviewed anyone's peanut butter and jelly sandwiches. So I knew there were no little Hineses.

Even with the heartiest of fancy meals costing upwards of three dollars, there was no mention in the listings of credit cards. To fill out a page, Mr. Hines would drop in an occasional comment from a reader who didn't have a full name: "Just returned from Florida. Used your book, which was a big help on trip.—Mrs. L.S.C., Iowa."

Because of his vast highway experience, Mr. Hines would also include some handy travel tips: "Books are helpful when traveling by bus, airplane, or train as well as by automobile. Useful when visiting unfamiliar localities or cities." He might let slip an understandable, though definitely not braggy, touch of advertising for his own writings: "They make appropriate gifts for birthdays, anniversaries, vacationists, travelers, holidays, and for party favors."

Being a person of the strictest integrity and since suing had not yet become a way of life, on occasion Mr. Hines was forced to include in his book a stern warning: "It is a distinct disappointment to me to learn that a surprising number of people have gone to listed places and received free meals and lodging because they have claimed to be relatives of mine," he wrote. "Please remember I have authorized no one to make such demands and they should be refused." In my childhood, anyone in a position to go around authorizing or not authorizing things required considerable respect.

Every year, Duncan Hines also said that he really wished it was possible to meet us, which was nice of him. "I think of myself in the role of host to many, many friends," he said. "And I hope someday soon, I may welcome you at my home and office." The book had a picture of his house. And then the page had his signature, just like the Smucker's labels. Some lower species might lie to a restaurant hostess to finagle a free meal. But no one would ever dare to sign their name to something false. I authorized myself to be impressed.

So you might understand why over all these years I have thought that "old Dunc," as my Dad called him, was a real person. Being a young backseat rider, it never occurred to me to wonder how one man could possibly eat so many meals at so many places every single year. Over time, I learned that Mr. Hines was a printing salesman traveling (he called it "motoring") all over the country back when U.S. 1 was young. To pass the time at speeds up to forty miles an hour and to save time the next trip through, he and his wife kept notes on where they ate (he called it "dining").

Of course, his friends knew of this habit. So they sought advice. A Chicago newspaper did a story on this hobby that showed how the American masses were beginning to creep out of their local cocoons. Strangers began to call Mr. Hines. In 1935, instead of the usual Christmas cards, Duncan and Florence Hines sent out a list of their favorite eating establishments. Within months, they were in the publishing business.

According to an obituary, which *The New York Times* obviously published by mistake back in 1959, Duncan Hines should be about 112 years old now. I know this because, although he doesn't go around reviewing restaurants much anymore, Duncan Hines does take his cookies and cake mixes around to every grocery store in which I've ever been.

Someday, I really will catch up. Someday, I'll come around the corner of the aisle from Pet Foods and Bottled Water and there he'll be, just beyond Spices but before Feminine Hygiene, a balding Duncan Hines in a clean apron signing boxes of his cake mixes and putting them up on the shelves.

I know Duncan Hines will be short. So he'll look up. He'll smile. And he'll say, "Hello, Andy. I see you made it." We'll shake hands and I'll ask him how he ever got his updated listings into our glove compartment in time for the Schenectady trip. We might even go back to his home, where Florence is scooping the cake mix into boxes for her husband to take to stores after lunch. She'll want to brew tea, which I can skip, and serve soft cookies, which I can handle. Mrs. Hines won't have one of those microwave ovens. But she'll have a cake mix for it, because you know how those folks are in Cincinnati, the ones at Procter & Gamble, who are so fond of the word *Instant!*

Depending on how frail Mr. Hines seems, I might also tell him how we never did get to the Ocean Sea Grill in Bridgeport, Connecticut, which he probably knows already. But that wasn't his fault.

You see, if it didn't walk or fly or come out of the ground, my family didn't eat it. And judging by Mr. Hines's listing from 1946, the Ocean Sea Grill on U.S. 1 would seem to specialize in things that swim:

BRIDGEPORT, CONN.　Ocean Sea Grill

1328 Main St. Open all year, noon to 9:30 P.M. Closed Mondays. ACond. Fresh sea food cooked to order and properly seasoned. Specialities: stuffed and broiled lobsters, baked filet of sole with lobster sauce. L., 85c to $1.10; D., $1.10 to $1.75. Also à la carte. L.

"No," said the operator. "I have no listing for an Ocean Sea Grill at 1328 Main Street."

I knew it. I was going through a surviving Duncan Hines guide (show me one person who can toss even an aged Bible out in the garbage) and after nearly a half-century, there were no signs of life in these old places.

"Wait," said the operator. "I do have an Ocean Sea Grill at Seventy Main Street. Here's the number." And she authorized the machine to talk to me: "three three six two one three two."

"No. No. No," said Bob Rolleri. "We have steaks, too. And chops. Italian chicken. And a new dish, lobster-shrimp diablo."

I asked him if he knew about Duncan Hines.

"Yeah, sure, I know him," said Bob.

"You do?! Where is he? When does he come in?"

"No, I mean, I know about Duncan Hines. My father told me about him."

Bob's father, Emil Rolleri, was the son of Eugene Rolleri, founder of the restaurant back in 1934. They moved the place in the forties so they can seat 225, which they do most Saturdays, instead of 75, which they can get some Mondays. They are closed Sundays now; people don't seem inclined to go out for dinner that day as they used to in downtown Bridgeport. In fact, a lot of people have moved out to the suburbs and don't ever come downtown anymore. The city hall neighborhood is not the mecca that it once was for law-abiding citizens, especially after dark. Even the route of U.S. 1 was rejiggered years ago to avoid the downtown.

Sometimes it seems as if the only thing they're building in the neighborhood is more prison. Of course, the Palace and Majestic vaudeville theaters are closed across the street. Built in 1915 atop an artificial pond where fans blew air up into the theater for a crude kind of air conditioning, the old Loew's theaters have been abandoned for twenty years, like many nearby buildings. A developer tried to tear them down for a senior citizens home, but the usual zoning fight erupted, so that's apparently dead. Now there's talk someone might restore the Palace and Majestic to legitimate theaters. But that takes a lot of money and even more confidence in the future of old downtowns.

But the Ocean Sea Grill still packs them in, though only 30 percent come from Bridgeport now. The rest live in nearby towns and suburbs. Bob has customers who have grown up with him; they were youngsters when their parents took them out to eat at the little restaurant run by Bobby's father. One customer the other day even brought in a menu from 1943, which his parents had kept as a souvenir in a desk drawer. Of course, the prices have changed. Lunches go from $5 up to $20 and dinners from $9.50 to $23.50. And even though there's no more after-theater crowd, the grill stays open longer, 11:30 to 10:30, to squeeze in a little more business—young couples who are both working late, those kind of people.

Bob has an idea to put a cooperative lobster plant right next to the restaurant as the kind of theme attraction that people seem to like more nowadays. They could eat it at one end after watching the creatures come in the other. But it's all based on serving good food with a personal touch. "It changes and it doesn't," said Bob. "You know what I mean?"

PART II

NEW YORK TO WASHINGTON, D.C.

The City, Gotham, the Big Apple, Where It's At, Man

AT FIRST GLANCE, U.S. 1 in New York is not much to look at. Second glance, too. The minute after the first road opened, the first roadside food stand opened. Many others followed. Even old Dunc could update his restaurant guide with ease:

BOSTON POST ROAD, N.Y. McDonald's

Everywhere along both sides of every road. Open to any vehicle. Open all year, 6:00 A.M. until ? ACond. Freshly thawed meat patties cooked before you left home. A taste for condiments is advisable. Specialities: french fries, Coca-Cola–flavored carbonated water, salads from New Jersey truck farms. B., L., & D. à la carte. Proper dress req'd, also ability to repeat order several times to an uncomprehending teen.

MAMARONECK, N.Y.   McDonald's

See Boston Post Road.

LARCHMONT, N.Y.   McDonald's

See Mamaroneck.

Motorists on 1 cross the Connecticut–New York frontier at a virtually unmarked community named Byram, because that's where the Indians used to go to buy rum (really). Today, the road is almost as good as it was back then. The first community of any size is Dominic Brigante's old hometown, Port Chester, where the sign for the NATIO AL TYPEWRITER S RVICE needs service itself. If the traffic isn't too bad or halted because the fire engine is departing, drivers pass a commercial chronology of modern urban America, with the store signs yielding their origin like layers of dirt in an archaeological dig—Moy's Laundry, Luis' Place, Scandinavian Furniture, Someone's Clutter Antiques, El Carmelo Mkt., Jack's Fabrics.

At 3:00 P.M. every day, everyone in town seems to be wearing white; the day shift nurses from the Port Chester Nursing Home are all heading for the bus stop and home.

The shifts have ended at the old Life Saver factory. For many generations, shift after shift turned out the multicolored candy circles that kept youngsters in heaven and dentists in business. They were invented in 1912 by Clarence Crane, a Cleveland chocolatier who decided to compete with more-famous European mints but first had to overcome Cleveland's summertime heat. Crane hired a pill maker and told him to design a shape for his new hard candy that was different from the traditional little pillow globule. The original ad slogan: "For that stormy breath." For one entire generation, I consumed them like, well, candy. I thought Life Savers were produced in my mother's purse. It sat there on our car trips and every once in a while she'd say, "Anyone want a Life Saver?"

"I do!" I'd yell, leaping up from the backseat. But Dad already had his right hand out. Dad always got the first Life Saver in the pack; that's what being a dad entitles you to. Trouble was, he always got the red one. No matter which end she opened first with that waxy little thread, there was a red Life Saver. And, zip, it went right into Dad's mouth. I always got the orange or, worse yet, green ones. And Mom did not allow rearrangement of Life Saver packs to remove the red ones and pack all the green ones on the bottom to melt and rot in her purse over the winter. No, sir. Taking the next little Life Saver from the pack, regardless of flavor, was the best kind of practice for learning how to take life's blows. Take 'em as they come, one after the other, as the Fates decide. Play the cards you're dealt. And, No! You may not open *both* ends of the Life Saver pack.

When I was slightly older, I pictured old ladies in babushkas packing the little rolls of candy and conferring in coded language on which end they thought Andy might open first. When I had my own pack of Life Savers, I'd debate that momentous issue with myself, usually leaving it to the usually infallible Eenie-Meenie-Mynie-Mo system of selection. Inevitably, I was wrong. Pick this end, the red one was on the other. Pick the other end, no, it was on this one. It still is. (This puzzling pattern did, however, enable me to appear polite at no real cost; if my parents were watching, I could offer any nearby friends a Life Saver and work my way down more quickly to the cherry level, when I alertly took my turn.)

And Dad didn't suck his Life Savers, either, like you were supposed to. Oh, no. He ate them! Just crunched them right up, one after the other, like popcorn. Hand out. Life Saver in. Life Saver up. Crunch! Life Saver gone. Hand out.

Wincing, I'd barely get my green one wet, and his right hand was back out for another Life Saver, probably lemon, which was pretty good, actually. I'd suck away and suck away sourly. And his hand would be out again. But finally I was done. I ate my green beans. Now, gimme dessert. I'd put my hand out and—oh, the mysterious, possible-lottery-winning anticipation of it all—orange. The next one was orange. Blah. Always the *next* one was cherry. And Dad crunched it. He didn't even have time to taste it.

"You got the cherry one again," I whined.

"Oh, yeah?" he said. See. Red ones were wasted on him. It could have been a cheese-flavored Life Saver, he wouldn't have known.

I complained about this on every trip. "You chew them too fast," I said.

"This way I get more," he replied.

How brazen! I should call his mother. Then I got a brainstorm. Sheer genius. "If you keep chewing them so fast, you'll get an upset stomach."

Proudly, I looked over at Mom. She was always very concerned about such things. I pointed to Dad. "Eating candy too fast." Nothing. No response. She didn't take away his cherry Life Saver privileges, not even for one turn. She just gave him another one. (Perhaps we were into the Butter Rum pack by now.)

"This isn't fair!" I said.

"You eat your Life Savers your way," he replied, smiling. "And I'll eat my Life Savers my way." A ruthless, ruthless man.

Now I know where Life Savers came from, Port Chester right along U.S. 1. Except they don't come from there anymore. In 1967, the Life Saver people needed a larger, more modern factory. (They had gotten my letter suggesting an all-cherry pack of Life Savers.) So like many manufacturers along U.S. 1, they left the old five-story factory on the two-lane highway and moved to Holland, Michigan, where each year now 150 workers turn out 7,486,761,000 Life Savers, not by hand. Profits from these little pocket packets of pleasure are helping to finance the largest leveraged buyout in American business history, the $25 billion takeover of Nabisco Brands and R. J. Reynolds Industries down in Winston-Salem by a bunch of New York financiers. For the little candy O's, the main impact was that its corporate parent is now a North Carolina cigarette maker, its corporate brother is Planters Peanuts, and its company name was changed from Life Saver to Planters Life Saver. Personally, I don't care as long as they don't add bran to Life Savers somehow. However, I do think all this billion-dollar business-borrowing and -buying may have less to do with corporate financial ambitions and more with various boards of directors full of dads fighting for first dibbies on the red Life Saver.

* * *

Beyond Port Chester, the grass at the Rye Golf Club is covered with cars on a sunny summer Sunday and the chemically balanced swimming pool is packed with humanity who opted against a dip in the nearby chemically imbalanced Long Island Sound. In Mamaroneck (which depending on your translator means either "where the fresh water falls into the salt" or "he who assembles the people"), other sun worshipers lounge on the grass, coveting the craft at the Post Road Boat Yard and the yacht club not far from where D. W. Griffith made the village an early center of moviemaking. Meanwhile on this sunny weekend, a lone elderly jogger makes the track rounds at the silent high school football field that, like so many others along U.S. 1, is so filled with high hopes and physical passion on a few fall Friday nights.

Once the county was a mere stagecoach stop on the road to somewhere else, a perfect setting for James Fenimore Cooper to write his stirring sagas. Travel on the Post Road was so uncertain in those days that many travelers wrote out their wills beforehand, just in case. Then, Westchester was a sprawling rural area where Peter Munro, a nephew of John Jay, built his manor house on a slight rise of ground. Bothered by the noisy bustle of Post Road even in the early 1800s, the presuburbanite had his Scottish gardener wall the house off from the road by planting a thick row of larches on the mont (Larchmont). Eventually, of course, many communities developed in Westchester—Harrison, Mount Vernon, Rye, and New Rochelle, which was founded by Huguenot settlers, including families with such future famous names as Faneuil and Revere, after the Edict of Nantes was revoked. Tom Paine, the Revolutionary author, lived in New Rochelle, too, when New York gave him an estate seized from a loyalist. Tom ended up being buried on his own property there because, as an agnostic, no churchyard would have his remains in their consecrated grounds.

These and many other communities that spread inland eventually became suburbs of New York City, affluent appendages where, thanks to the independence of individual transportation, millions could flee each night. Once, more than half the adults in these towns commuted into New York. Now barely a third do. Westchester has become a developed and developing urban area in its own right with all that that offers and entails, from a symphony and theme park to crack dens and aging welfare motels, where teens of another era once met to grope through some clandestine initial sexual encounters.

Larchmont's automobile alley, including dealers for Maserati and Alfa-Romeo, is one indicator of the affluence that has seeped through much of the county. And along 1 in these parts, the increasing use of these vehicles's horns, my favorite index of impatience, can mean only one thing: It ain't far to New York City.

The last gasp of scenery comes in Pelham, once the Pell estate, then an isolated community where Cato, a South Carolina slave who bought his freedom, became a Westchester restaurateur, whose okra soup, curried oysters, and roast duck once made his house a favored dining destination for parties of sleigh riders from New York. Washington's army retreated along here after the Battle of Pell's Point in October 1776. Today, Pelham is an unusual suburb, one of those old enough that the driveways still curve around to the back and the trees that line its tidy streets no longer require guy wires to hold them straight.

It's a good thing that so many of the nation's immigrants arrived in New York City by ship, sailing through the Verrazano Narrows and up the Hudson River past the Statue of Liberty and the Battery. If they came in the back door, along U.S. 1, they might not have stayed. Even our country's founding fathers could not stand New York as the capital for more than two years, preferring to reside instead in a swamp not far from Mr. Washington's Virginia manor.

U.S. 1 bursts out of sylvan Pelham, ducks under the (Anne) Hutchinson Parkway, and plunges into the raucous cement cacophony of the Bronx. Although most Manhattan residents do not know or acknowledge it, the Bronx is one of New York City's five boroughs. It is named for Jonas Bronck, who built the first manor house north of the Harlem River. The next year, 1639, the Dutch West India Company began developing a housing subdivision and called it Broncksland. Four years later, the Indians cornered Ms. Hutchinson (a fate worse than death, I believe) just down her parkway on a nearby point of land called Throgs Neck, now the site of a bridge to Long Island. Such edifices are concrete tributes to the memory of Emperor Robert Moses, who was never elected to any office but shaped the physical fate of the New York City area by his grand public construction schemes. One of his narrowly missed goals: build one parkway for every citizen.

About sixty years ago, a team of WPA writers found this Bronx section of 1 "dismal and uninteresting. Gas stations, auto junk yards, diners, and third-rate roadhouses line the highway. Weedy lots sprawl to the backdoors of private homes and apartment houses." Today, much has changed. There are fewer weedy lots and diners, more gas stations, and the purveyors of junk are called salvage yards.

With I-95 and U.S. 1 paralleling each other an easy sniper shot apart, the street is pocked with potholes and lined with chain-link fence topped with razor wire. Many establishments like the Top Tomato Mini-Mall go for vast sheets of corrugated steel shutters, which are rolled down to cover the entire storefront in off-hours and present a colorful tableau of gang graffiti. Whining jets circle overhead bound for La Guardia Airport. Many vehicles are double-parked while their drivers perform repairs (assuming the cars belong to them), chat with friends or business associates on the sidewalk, or argue with others. In front of a fading mall, where a handful of trees try to survive in the withering heat, large crowds of shoppers, some wearing shirts, stand in line for the blue-fuming buses, clasping a bulging handful of hard-earned change. The bus fare is $1.15 or more—exact change or token only; United States dollar bills are not recognized as legal tender by bus drivers of the New York City Transit Authority.

Towering above the neighborhood over by the vast sales depot for school and prison buses is Co-op City, a battalion of apartment buildings up to thirty-three-stories tall, so similar that through the windows, lighted ceiling lamps all hang atop one another in exactly the same place, five hundred feet deep. At major intersections, gangs of youngsters swarm up to drivers' windows, hawking everything from newspapers to handfuls of car deodorizers at prices hastily negotiated before the light changes.

The Bronx is home to more than a million people. But to outsiders, it seems an urban free-fire zone where anyone unfortunate enough to have auto trouble returns with a tow truck to find a few wheels missing. Mel Clark, the New Hampshire used-truck dealer, tells the story of witnessing a traffic accident during one trip. A car had struck a motorcyclist. Arriving on the scene, sirens blaring, police could not understand why the injured pedestrian was wearing a helmet; someone had already stolen the motorcycle. Like other urban American battlefields, New York has a problem with auto thefts. Every year now seven vehicles are stolen and another four nearly stolen for every thousand vehicles on the road. And while it's hard to hide a car, only six of ten are ever recovered. The others end up as taxis in Haiti or somewhere or, more profitably, as pieces sold on the auto-parts black market, according to the perverse economics of today that make vehicles worth more disassembled than whole.

New York City also has a major problem with abandoned cars—plateless, rusting hulks cannibalized by vandals and scavengers and left to deface virtually every neighborhood and feed the flammable fears of city residents. The average year finds 140,000 (!!!) such car carcasses consuming priceless parking spots.

But New York City has a tough new program to crack down on this urban blight: officials issue tickets to 14 percent of the violators, an enforcement program that no doubt has the other 86 percent quaking in their dirty Reeboks.

Like a suburbanite who accidentally exits the expressway to become lost in a labyrinth of streets with missing signs, U.S. 1 hurries through the city out of town toward the bridge, its windows locked, unnoticed and happy for it. Just before the two-tier George Washington Bridge link with New Jersey, U.S. 1 briefly joins its old bridge buddy, I-95, to form the Cross-Bronx Expressway. Even on a bright weekend afternoon, however, *expressway* is a misnomer. The problem is that after many years of neglect, New York–area bridges have begun to crumble, literally—pieces falling off, steel planks being one bolt away from plunging into the river. So repair crews have shut down a couple of lanes, meaning that the motorized traffic with all of its computer chips is creeping along at one-half the speed of an olde stagecoach.

In the sunken alley that is the modern expressway, the refuse whips around by the ton, obscuring the city's official signs threatening a $250 fine for littering. In such eight-lane parking lots, idle ears tune in radio ads for an effective treatment for head lice (unhatched eggs and all) and idle eyes light upon a large sign painted on a brick wall: ARMORY DOG AND CAT HOSPITAL DR. ALBERT BURCHMAN, VET. SE HABLA ESPAÑOL.

"In the country," said Dr. Burchman, "people use animals for work. In the suburbs, animals are pets. In the city, animals are protection. We don't see a lot of parrots in this practice."

Studies show pets ease emotional isolation among young and old. Petting a cat has been shown to reduce blood pressure (in the human). Expenditures on pets show how large an attachment people have to these creatures. Nationally, cats are by far the most popular pet; in fact, Americans buy more food for their 58 million cats than for their babies. They spend $7.5 billion annually on food for their 109 million dogs and cats, making pet chow by far the largest category of packaged food (and gourmet pet chow a mushrooming market niche). They also spend an additional $6 billion-plus on pet medical care, most of it uninsured and an increasing proportion of it mirroring human medicine with specialized practices, sophisticated technology, and medical machines.

This is all good news for the country's 50,000 veterinarians, a hundred or so of whom have offices along U.S. 1 and many more of whom nowadays are women. One of them, Dr. Larry Berkwitt, in Darien, Connecticut, sees households treating their dogs and cats as another full member of the family and willing to provide as much medical care as they can afford. And despite the manpower complications, he must also run a boarding kennel. "It's important to your clients," he says. "They want to leave their pet somewhere they trust."

Without house calls, Dr. Berkwitt sees perhaps ten thousand animals in a year of twelve-hour days. And he's seen some changing patterns in pet ownership on the shoulders of U.S. 1. Once, mainly families were owners. Now, many singles seem to need the relationship. "The young guys, say, eighteen to twenty-five," he says, "they don't like itty-bitty dogs. They go for the tough ones, the rottweilers and shepherds. It's the elderly who go for the shelties and Chihuahuas."

But not in New York, where some city kids grow up thinking that sparrows grow up to become pigeons. "People here are so afraid of crime," said Dr. Burchman, who keeps a Doberman around his office. "People in this area tend to go for your basic vicious pet. And then they name it Tyson or Rambo." This can create a challenge if the pet fails to understand that the doctor is here to help his owie. "I've got customers whose pets won't let their owners back in the apartment," adds the doctor. "The owners love it. It makes them feel safer." He also has pets who show up with gunshot wounds from their job of protecting an apartment.

What's the most popular kind of cuddly dog in Dr. Burchman's Bronx practice? "No contest," he said. "Pit bulls."

First Place Not Named ——port

ON OCTOBER 24, 1934, President Franklin Roosevelt, former governor of New York, dedicated the George Washington Bridge across the Hudson River, and the route of U.S. 1 changed, too. Previously, it ran down through Manhattan as Third Avenue and crossed over to New Jersey through the Holland Tunnel. In an unusual example of public foresight, the new bridge was built with 105,896 steel cables, enough to support a second lower deck on the off-chance that traffic increased in future years.

Although it is death-defying to sneak a peek amid all that angry traffic, the view from the bridge on a clear day is one of the best in the East—the endless apartments of upper Manhattan, the towering skyscrapers of lower Manhattan, the white dots of sailboats bobbing about in the immense river so far below, and the sheer cliffs of New Jersey's Palisades ahead, much of them heavily forested. From this height, the river water looks a beautiful blue.

It is not hard to imagine General Washington standing on the Jersey bluffs in Fort Lee (another general) and guarding the river in 1776. The plan was to keep the British fleet from sailing up the Hudson. The plan didn't work, however. And after one of the nearly one hundred Revolutionary War battles played out on New Jersey soil, Washington and his men were forced to retreat again along the highway. But as an example of the kind of golden luck that colored that Revolutionary War and others later, because it was not a crushing defeat, it became a victory for the amateur warriors. So even an American defeat prompted increased French aid and British demoralization.

Prehistorically, the Fort Lee area was a dinosaur pathway and playground (later, it became the site of a famous amusement park for humans). But for long decades, Fort Lee was a sleepy small city whose slow-paced daily life was punctuated during the Civil War by Union ironclads using the Palisades for target practice. The community's principal economic claim was production of Belgian paving blocks to line New York City streets and docks, the stones being blasted from cliffs below the present bridge site. A movement led by an alliance of women's clubs halted the blasting at the end of the nineteenth century. A park commission was formed, and the American ecology movement had one of its early victories.

For fifteen silent years before talkies, Fort Lee was the capital of American moviemaking; many classics were made there, including *Les Miserables*. And the city suffered numerous serious fires from the volatile nitrate film, until cinema moguls decided they needed Los Angeles's year-round warmth to sit around swimming pools and do lunch.

There are beautiful parts to the nation's most heavily urbanized state. New Jersey has tidy neighborhoods, mossy woods, isolated lakes stuffed with bass, pristine waters draining semiwild mountain ranges. But none of them are along U.S. 1. New Jersey's unenviable reputation for dirt and grime, fumes and traffic, and the general ugliness and rudeness that accompanies urban industrialization is based on the aging monochrome world lining its transportation corridors, the nation's most lethal.

The eighty-mile section of U.S. 1 that runs from the Washington Bridge (known as "the G.W." on the area's incomprehensible traffic reports) down to the capital at Trenton is the state's oldest transport corridor—also the newest and most congested. In the North, it consists of mile after mile of crumbling pavement, divided by narrow cement barriers, cramped by overgrown sidewalks, and lined with warehouses, gas stations, truck depots, and motels that do not openly advertise hourly rates. On the brief but grim 3.4 mile Pulaski Skyway, which is intended to honor a Polish nobleman killed in fighting for the American Revolution, vehicle wheels crunch the glass from last year's accidents. An elevated U.S. 1 passes by an oily industrial vista of second- and third-floor windows belonging to old-fashioned factories, before vaulting across vast bogs, fields of overgrown swamp grass, and dead-end dirt lanes that make ideal dumping grounds for the used bodies of former mob members.

U.S. 1 in New Jersey avoids the downtown of major cities because it was thought travelers could make better time that way, which is true in a major metropolitan corridor until everyone else takes the same route. The highway ducks past some less-than-scenic refineries and Newark International Airport, built on more landfill. It cuts across Union County, which the Dutch settled in 1609 but left on British request in 1664. The highway slices through one side of Elizabeth, which is conveniently located just five hundred feet west of New York City (a modest and often polluted waterway is the only gap from Staten Island, which is trying to secede from New York City). Elizabeth was where the about-to-be president Washington left his horse to catch the barge to his New York inauguration. And it is close to the tavern where an atheist named Harper was celebrating the closing on his new house when the famous tornado of 1835 struck. According to local legend, Mr. Harper ran into the street and defied God to kill him. A large wooden beam promptly crushed him, which prompted the area to be severely pious for several years.

As it silently creeps by private homes and stores and ancient cemeteries, whose stone markers are blackened and toppled, U.S. 1 cannot seem to decide whether it wants to be a major thoroughfare or a local street. So it tries both. It carries shoppers going two miles and workers going ten and trucks going one hundred. It serves students at two of New Jersey's best-known universities, Rutgers and Princeton, which late last century happened to play the country's first intercollegiate football game

(Princeton lost, of course, six to four). Paul Armstrong drives on 1 to see state legislators in Trenton. With high-power lines draped overhead, the road has direct driveway access to topless nightclubs and vegetable stands, to the corporate headquarters of Dow Jones and Johnson & Johnson, and to roadside vendors offering an impressive array of velvet Elvis paintings. The highway serves a diner named for Queen Elizabeth, who is not known to have a financial interest in New Jersey eateries, as well as Macy's, Gary Alexander's Self-Defense Studio, countless "garden" apartments, and Ford's vast, windowless Metuchen assembly plant.

Forty-eight miles from the George Washington Bridge ramp, it also provides access (southbound, anyway) to the U.S. 1 Flea Market, an immense collection of independent merchants who gather under one roof every Friday, Saturday, and Sunday to offer inexpensive goods to thousands of tourist shoppers. "You get really good crowds here," says Lou Bianchini, who peddles toys and leather jackets along the far aisle. There, under the heading of bazaar bargains, shoppers can get everything from laundry soap and Batman costumes to sausages and tattoos, from glittering sweatshirts extolling the virtues of tropical Puerto Rico to backless formal gowns, from gold chains and discount liquor to wild birds and goldlike chains. They also sell discreet lapel buttons for the modern woman that read, "Stop Staring at My Tits."

The market also offers country-western music bands. During one break in the music the other weekend, a large crowd gathered around to watch someone's car go up in flames. Within moments, the New Brunswick fire trucks were within one hundred yards of the ten-foot flames and tower of oily black smoke. But due to the cement barriers that divide 1 in that stretch of busy highway, it was another ten minutes before the trucks could proceed far enough north to make a U-turn and scream back south to the flea market's parking lot. And by then, as some spectators applauded and others dribbled back into the bazaar, most of the car had been consumed.

* * *

Just down the road is another driveway that Carl Schaab takes every morning into a nondescript factory that carries on the tradition of a nineteenth-century Jersey boy's unusual merchandising idea. After his birth, Aaron Montgomery Ward's family moved to the Midwest, where he grew up to become a traveling salesman of dry goods in 1865. Ward noted that unlike Europe, where peasants mostly lived in towns and commuted out to their farms each day, American farmers were scattered by the millions in houses all over the countryside. This enforced isolation was, in effect, mandated by Federal homesteading regulations, which denied title to the land unless the owner actually lived on the property for five years. But it created an entrepreneurial niche for someone who could bring merchandise to the buyer and unite this market through the mail, not accidentally eliminating the middleman store owner in town. Thus, with $1,600 of Ward's savings, $800 borrowed from a friend, and a one-page circular with 163 items, the business of direct-mail marketing was launched.

Today, direct-mail is a $183 billion-a-year business in the United States. Last year, 92 million Americans responded to pitches by direct-mailers, whose third-class mailings account for 39 percent of all postal volume. That is a colossal 4 million tons. This volume includes more than 116 million catalogues from L. L. Bean up 1 in Freeport. And this freight along U.S. 1 also includes more than 500 million little pieces of paper from Webcraft Technologies, Inc.

Webcraft and Mr. Schaab, a project engineer, have perfected the fragrance strip, those specially treated pieces of smelly little envelopes that make your magazine and department-store bill smell like the back of someone's ear. To some, fragrance strips with their attached order forms are an intrusive marketing ploy. But they do cater to two powerful American drives: the desperate desire to smell good and the perceived passion for saving time. As a result, fragrance strips have breathed new life into the $4 billion-a-year American fragrance industry, not to mention what they are doing to drive people crazy for vanilla, liquor, golfing, and even the leather seats in a stately Rolls-Royce. On the horizon are fragrance strips smelling like popcorn and lasagna and root beer. Yum.

Fragrance strips are actually chemical descendants of the scratch-and-sniff strips that appeared in children's books years ago before helping to sell deodorants and shampoos. But, it seems, women, who also buy more than half of all men's after-shave and cologne, do not like scratching things with their nails. Like so many of mankind's pioneering discoveries, the ability to make paper smell like something for sale—without scratching—was an accident.

Back in the 1940s, a research chemist for NCR Corporation was trying to find a way to prevent purple fingers, the stains left behind when changing his company's cash-register tapes. His solution—tiny capsules of ink safely sealed in chemical bubbles—became carbonless paper. It's not a great intellectual leap to envision these chemical pimples substituting scent for ink. When two impregnated pieces of paper are pulled apart, 100,000 little bubbles burst, releasing the scent and theoretically compelling readers to become buyers. In his lab, Mr. Schaab is working on all kinds of new applications; he now has a perfect fake-pine spray, which might go on artificial Christmas trees or in stuffy closets.

For much of every year on U.S. 1, Webcraft, which also prints the lottery tickets for several states, has six high-speed presses devoted full-time to producing fragrance strips every day around the clock. Giant rolls of special paper are fed into one end of the press. In a matter of seconds, the paper is printed, twisted, glued, folded, looped, glued again, coated with the secret formula, scent-bearing chemical slurry, sealed, and spit out the other end in piles for mailing or binding into magazines.

The trucks pull away from the loading dock's door and join the endless traffic creeping along the old stage-coach road. The problem is that planners did not anticipate suburbia as anything other than a satellite; many still don't. People would go into the big city in the morning and back out at night, and the roads and tracks could be built permanently that way like spokes. No one figured

so many people and companies would forsake the old center for living *and* working to create scores of new economic orbits with their own centers and overlapping centripetal and centrifugal forces. The number of New Jersey workers who now live and work in nearby communities once called suburbs is passing the 40 percent level and rising.

A twenty-six-mile stretch from New Brunswick past Princeton down to Trenton is a compelling and revealing example of the evolving new face of suburban America. Until very recent years, the area was known for its sod farms, vast fields of deep green that were mechanically sliced from the fertile topsoil below, rolled up, and trucked elsewhere to provide instant landscaping lushness.

This area now is no longer rural, no longer just a zoning variance, and not yet a separate city. The emergence of this stretch of 1 as a major business corridor dates from the early 1970s when Princeton University, anxious to control development of its surrounding area, protectively purchased a large tract of adjoining land. Called Forrestal Center, these 2,238 acres evolved into a major campus-style office park with nearly three dozen buildings, including two hotels, a conference center, apartments, and a shopping complex. This has, in turn, attracted numerous additional developments eager to rub digits with the prestigious 08540 Princeton Zip Code. Long, yellow earth-movers parked in roadside fields just recently farmed foretell of topsoil stripping next spring and occupancy not too many months later.

It is an ideal location for offices, halfway between New York and Philadelphia, and Washington and Boston. Experts expect the working population along the shoulders of U.S. 1 there to equal that of a major city in a few years. Yet very little housing is handy nearby, meaning most must travel from somewhere else and return there each night. Driving statistics show the same phenomenon nationally, with the average car trip getting shorter while the average travel time lengthens. So the highway is jammed at rush hours, which have become more rush periods. With traffic counts at individual corners increasing by the day, state planners have identified more than $750 million in improvements necessary on Route 1 and adjacent local roads just to keep them at the congestion level of the mid-eighties. Costs have gotten astronomical, with one lane of widened road running at $5 million a mile and a simple overpass to eliminate a single intersection costing $20 million. Some companies have introduced different work times to stagger the rush. The state encourages car pools and company vans, which is like using a sponge to tidy up the Johnstown flood. But, as usual, planning is fragmented among adjacent communities, with each approving the next development without regard to its regional impact.

"The biggest challenge we all have," one New Jersey county planner, George Ververides, told a reporter, "is trying to break up the marriage of people and the single-occupancy vehicle. We feel we have to have a lot of patience."

Good luck, George.

Serving the Nation's Would-Be Capital Since 1767

NOBODY NEEDS TO THROW a dollar across the Delaware at Trenton today. The bridge toll from the cement gully of U.S. 1 out of New Jersey's capital (William Trent) into Pennsylvania is only a quarter these days. Motorists can drop the exact change—the only toll the entire length of southbound U.S. 1—into a worn plastic basket without having to say "Hello" or "Good Morning" to a human attendant. A machine reads the coins and turns on a green light when you may go. (And Washington was going the other way, anyway, when he surprised the Hessians north of town.)

Morrisville is the first town encountered in the Keystone State (Pennsylvania is not named for the famous motor oil; its geographic position in the middle of the original thirteen colonies was considered crucial and Philadelphia was the national capital for twenty-one years, although this has not helped the Eagles in the play-offs). Morrisville, is named for Robert Morris, who, like Yale, Bowdoin, and the others, earned this distinction because he had a lot of money and gave away much of it; in this case, to help finance the American Revolution. In gratitude, Morris was made a signer of the Declaration of Independence, although another signer's name was cho-

sen for the famous slogan for signing documents, because it sounded funny to say, "Put your Robert Morris on the dotted line."

The unusual idea of placing a governmental capital roughly in the middle of the people being governed (instead of, say, near the king's favorite castle) is a democratic one that gained real momentum in more western states and county seats, places that New Englanders consider the Far West, like Ohio and Indiana. But on October 7, 1783, the new United States Congress came within two votes of making Morrisville the nation's permanent capital. Unfortunately for Morrisville's Chamber of Commerce and for Mr. Morris himself, who kept an estate of nearly five square miles in the area, George Washington did not like the idea of traveling so far from Virginia, even up 1. Southern interests favored Annapolis. And Alexander Hamilton did a deal with Thomas Jefferson and some real estate men farther south along the Potomac River closer to the big boys' homes. So Morrisville lost out on the capital idea.

But it did win the Moon business. Mahlon Moon made the first sale from his nursery—eighty-four apple trees for a shilling apiece—in 1767 from property he got from William Penn. Probably not anticipating that their land was in the path of a future shopping mall but simply seeking better soil, the Moon family moved its nursery operations in 1848 up U.S. 1 a piece to its present location. Now, the one hundred acres are surrounded by office buildings and apartments—Morrisville is a good commute to Philadelphia and back into New Jersey. And the real threat to the Quaker family's 225-year-old business is the road that brought them there.

Five times this century alone, authorities have condemned pieces of the Snipes nursery (in 1912, Hannah's grandma married a Snipes) to enlarge the lanes. With suburbia still growing as affluent families flee the financial costs and social troubles of the cities, there is no guarantee that the nursery won't be further nibbled away.

Even during the very tough times when the demand for wholesale nursery stock was not overwhelming, the family stuck together, with members pitching in to work and brothers sharing the property-tax burden, even absent any income. And while general manager Hannah Snipes says the family has produced a few too few males to help with the management and physical work, over the years Snipes Farm & Nursery has survived by its reliability and diversification.

In 1958, the nursery added to its wholesale landscaping business a picnic table for the excursion market, along with a retail houseplant, garden, and fruit business. This tapped into the developing do-it-yourself market. But that has stumbled somewhat in recent times for two main reasons: the growing number of rental apartments where residents do not care for their common gardens, and two-income couples who find less time to garden together but more money to pay someone else to do it.

I remember two generations ago my mother and father lavishing much care on the minuscule grounds that came with our floor of the rented house. In those days, *rented* meant "It's ours for a while, so we'd better take good care of it or else our reputation will be hurt" and not so much "Hey, this whole place belongs to a limited partnership of doctors and I pay plenty for them to keep

it up." Most summer weekends, there was something important for my parents to do to the large rock garden out front, which was puzzling to me because there seemed to be enough rocks around without growing more.

My mother had been born in the city but raised later in the country, where it was up to landowners, not the government, to keep things tidy. My father had been raised on a hardscrabble farm, where duties were set by the needs of the livestock and the rhythms of the seasons on the land. It was a grinding routine, I gather, but one that instilled a powerful sense of self-discipline and stewardship of the land.

As soon as we got enough money, my parents bought a house and small piece of property so far out in the country that no one at the downtown department-store charge counters had ever heard of our town, and it took Dad forty-five minutes to drive to his office. Today, thanks to a network of expressways, it still takes forty-five minutes to make the same trip and it doesn't matter whether the clerk at the department-store branch in the nearby mall knows her local geography, because it's the computer that needs to read the charge cards. Our house, which had been an old roadside inn at one time, got some work. But it was the surrounding property that drew my parents' attention. And it provided, then, not only scenery and satisfaction but crops of flowers and fruits and vegetables, according to the right season. The three of us provided the sweat. And my horse provided the manure—boy, did he provide the manure.

We burned our garbage or turned it into compost. The milkman took the milk bottles back. We returned the soft-drink bottles to be used again. Plastic was something cheap that came from Japan. Raspberries grew wild all over, and to enjoy that bounty, we wrestled with the mosquitoes who also lived there.

Today, berry picking is a do-it-yourself pastime on non-mall weekends for families in minivans seeking an unusual afternoon activity together (as long as this week's big game isn't on TV). Snipes has a few regular women who pick their own crops for the steamy satisfaction of canning at home. But the nursery relies heavily on the business of suburbanite families for the success of its do-it-yourself strawberry-, raspberry-, blackberry-, blueberry-, and peach-picking operation. Come fall, thousands of customers pick their own apples for sixty cents a pound, an estimated 60,000 pounds' worth per season, plus another 70 tons of pick-your-own pumpkins, and hundreds of cut-your-own Christmas trees.

Snipes sells homemade cider, candied apples, and other fruits and vegetables. Caravans of school buses start appearing in September to show nearly ten thousand kids what a farm hayride is. Some people even buy used cornstalks for autumn porch decorations. And what doesn't get sold gets mulched; the landscaping side sells one thousand tons of mulch a year, plus a quarter of a million trees and shrubs, gardening supplies, fountains, patio furniture, and there is a Christmas shop. At peak times, the operation has 180 employees, including enough Snipes family members to keep it going, for now, anyway. They used to have a hunt farm with horse shows, sales, and boarding, plus horseback-riding lessons. But that closed when Anne, a sister, moved to California.

To put some extra land to use, Snipes also offers a fifty-tee golf-driving range, a pro shop, two miniature golf courses, and an eighteen-hole par-three pitch-and-putt course—but no video arcades or midway machines. "We're for families," says Hannah proudly. "We're clean, wholesome, and have no mechanical rides."

From the nursery, U.S. 1 wanders through Langhorne (LAFAYETTE HAD HIS BANDAGE CHANGED HERE) into the Philadelphia metropolitan area under various names—among them, Martin Luther King, Jr., Memorial Highway, City Line Avenue, and the Roosevelt Expressway, where because Philadelphia's fiscal problems are so serious, the fine for littering is fifty dollars greater than New York's.

Along the way, the road, at that point a broad boulevard, passes another familiar name, Whitman—Stephen, as in Whitman's Samplers. Yes, after all these miles of pretty woods and lovely shores and ugly stores, I have finally found a full-fledged, unadulterated chocolate factory. The recent Willy Wonka of the 500,000-square-foot factory was Jay Shoemaker, a thirty-nine-year-old former production manager for Cheerios, who ate that cereal growing up but thought then that Walt Whitman, the poet, was also into candy big-time.

The factory moved out to U.S. 1 in 1961, 119 years after Stephen Whitman, "confectioner and fruiterer" (his friends called him Mr. Whitman), opened a small candy store downtown near the Philadelphia waterfront. France had the corner on the fine candy reputation in those days. To talk about Americans making candy back then would be like saying Americans make television sets today.

But Whitman made friends with the sailors in his neighborhood, and they began buying overseas and bringing back to him the rare but necessary ingredients to make his candies unusual. He produced one special package titled "Choice Mixed Sugar Plums." Later boxes were called "Pink of Perfection" and "The Fussy Package for Fastidious Folks." Affluent chocoholics, especially from Philadelphia's developing stylish Main Line, would peruse the racks of yummies in the Whitman store and assemble their own chocolate arrangements. Each regular customer's preferred mix of candies was kept on a separate card, so that he or she could simply order "my box." The Whitman reputation spread and he began shipping orders on the numerous stage lines that fanned out from Philadelphia like the spokes on their own wheels.

In 1912, according to company legend, the president, Walter P. Sharp, got the idea to put a standard mix of Whitman's chocolates in an old-fashioned-looking box decorated with an embroidered sampler like his grandmother made. The boxes were distributed through a developing network of drugstores. *Et voilà,* big business.

Saywell's Drug Store in my small town had an entire half of a shelf devoted to boxes of Whitman's chocolates, which always had a surplus of caramels and never enough white creams to suit my fancy. But the rigid boxes did seem to please generations of mommies on Mother's Day. "He picked it out himself," the fathers would say, as if shopping for Whitman's involved a lot of skill and any thought beyond pointing at a pile of yellow boxes looking precisely alike.

Today, the factory has 1,000 employees, 105 of whom do nothing but pack the precisely programmed pieces of the three-pound pack. Little robots pack the one-pounders, which, fortunately, was not the case during World War II when human packers stuck flirtatious notes into war-bound boxes; a number of marriages resulted. And like Snipes Nursery, Whitman's Chocolates has had to diversify.

Grinding their own beans from Africa and South America and brewing the thick chocolate in thousand-gallon vats with wooden paddles you'd love to lick clean for your mommy, Whitman's now makes two hundred different products. They include chocolate for other candy companies, candies in heart-shaped boxes for Valentine's Day, and, of course, Whitman's Lite with the requisite "reduced calorie" centers to reach that market niche that is overweight with guilt. Every year Whitman's sells 25 million pounds of chocolate, which comes out to about 1 billion pieces. That's a lot of chocolates to squeeze the bottom of so I can peek at the center.

Like many former family companies, Whitman's is now part of a conglomerate. It is owned by the Whitman Corporation, which is the new name for IC Industries, which was the new name for the former Illinois Central Railroad, which decided to start changing its name after it spun off the bothersome railroad that was its onetime reason for being and instead got into Pepsi bottlers, Pet milk, Midas Mufflers, and chocolates.

As Hershey's, another famous Pennsylvania chocolate maker, did for many years, Whitman's long disdained anything as crass as advertising. That required some mental adjustment by Mr. Shoemaker when he arrived from Cheerios and its $50 million ad budget that had drilled allegiance to that cereal into the minds of a generation of little "Lone Ranger" watchers. The feeling in the marketing department is that the typical Whitman customer—aged twenty-five to forty-four, married, with two kids—knows the Whitman's franchise after all these years or is more persuaded by word-of-mouth, convenience, and price. And although Christmas is the richest candy holiday, more than 40 percent of Whitman's Samplers are sold for birthdays, which are not confined to one season (despite the sexual rhythms of logging families up in Fort Kent). So how can you target those ads?

This is not to say Whitman's does not seek new ideas. The other day during her boss's daily plant tour, Lynn Paff-Connor approached him. "You know what we should do?" she said. Her idea was to design different boxes of candy suitable for giving like a greeting card on numerous occasions from anniversaries and illnesses to birthdays and christenings.

So Whitman's will.

Out in front of the factory, where Whitman's president jogs off any extra chocolate pounds, U.S. 1 runs down a broad grassy median that separates Charley's Pizzeria from Sheehy Ford from Pearle Eyeglasses (in case you suddenly develop the need for eyeglasses within about an hour). The highway used to cut through downtown Philadelphia, but now it ducks to the west to bow around the populous mass on what is popularly called City Line. Not too surprisingly, this shady old street is the line between the city and the suburbs. On City Line across from the Overbrook train station of the famous Philadelphia Main Line sits an old Sunoco gas station that hasn't sold fuel in a decade. Johnson's Flowers is using the easy access of a corner gas station to peddle blossoms.

Thanks to overnight international air freight, they sell orchids from Thailand, tulips from Holland, ginger from Jamaica, eucalyptus from Australia, bird of paradise from down 1 in Miami, and roses from Colombia. Lots of roses. Sometimes ten thousand roses in a holiday week, when Jenny Kirby, the ebullient manager, boosts a special on the movie-style marquee outside—ROSES $15.99 DOZ. Some dash in by car, easily park, quickly buy, and shortly leave. Those with simply no time to stop and shop can phone to have a stranger give thought to the floral arrangement for their loved one and then have the blossoms delivered. To save even more time, many fax in their order when leaving the office and it is ready upon their arrival.

Women are the most frequent customers. They're usually buying for themselves, although a growing number seem to feel comfortable sending flowers to men friends. Women customers come in the morning, when they are apparently thinking of such fresh things. The men, who are usually buying flowers for someone else, come in the evening, after the train, before the car, and during the guilt. Jenny thoughtfully advises them to cut the stems again at home and to do it underwater so no air bubbles erupt in the stem. Such blockages to moisture can cause premature drooping, you see, which no adult likes, especially grown men.

To encourage repeat customers, Johnson's has a Frequent Flower Club with an official punch card—one square for each five-dollar purchase. Ten squares punched earns a free dozen roses. "It keeps them coming back," she says.

There is a steady business, too, from the hospital around the corner and from St. Joseph's University, which prompts Johnson's to stock a lot of crimson and gray ribbon around prom and homecoming weekends. St. Joseph's exists today, nearly 150 years after its founding, in part because Pennsylvania's Quakers, unlike the Puritans up the road, tolerated Jesuits and other groups. Surrounded by a troubled city in 1927, the university moved to a beautiful leafy campus and built its neo-Gothic buildings as close as it could get to the developing suburbs—a sidewalk's width away—without technically leaving the city. St. Joseph's, the seventh-oldest of the country's twenty-eight Jesuit colleges and universities, can still see downtown Philadelphia from Hawk Hill, where the old carillon bells in the Barbellin Tower have been replaced by eight 360-watt speakers. They broadcast amplified simulated bell claps on the hour at less cost and maintenance than authentic clangers.

The school still draws students from both sides of the sidewalk. But like many American institutions of higher learning, St. Joseph's is in tight financial straits and has had to launch an unprecedented $60 million capital fund drive, which will determine much of its future.

I'm not suggesting this for university administrators. But earlier residents along 1 had an admittedly crude but nonetheless effective manner of expunging debts in Colonial days, at least as they related to widows seeking a new spouse. Which, come to think of it, is actually not yet much of a problem for Jesuits. A deceased husband's lingering debts were an understandable deterrence to any serious suitors. According to the practice, called a "shift marriage," the woman could exorcise her deceased husband's debts by running across the highway four times stark naked save for a small shift. Traveling the entire length of U.S. 1, I did not see any of these. But I imagine them being done at night and, if in winter, rather briskly. Mary Webb did it in 1725 for the love of Henry Jun. Sarah Stephenson did it for John Garrett. And Abigail Culverwell did it on February 22, 1720, inadvertently beginning the celebration of a future president's birthday, which is marked somewhat differently in our enlightened times. Today, of course, we honor our deceased first president by staging department-store sales.

Colonials apparently saw no contradiction between shift marriages and bathing laws. See, bathing was frowned on because it promoted nudity (unarguably), which could only lead down the path of sin and degradation to Promiscuity. So Pennsylvania and Virginia enacted laws to either ban or limit bathing. A Philadelphia ordinance threatened a jail term for anyone who washed more than once a month, which helps explain the rush to the suburbs around that city.

I do not know if General Washington invented the aphorism about losing the battle but winning the war. But he did collect another one of his defeats along the main road south of Philadelphia at the Battle of Brandywine. That's where Lafayette was wounded and not far from Valley Forge and the worldwide headquarters of

AAMCO Transmission service, the Delaware County Prison, Swarthmore College, and the office of Stanley Blazejewski (pronounced Blazejewski).

Dr. Blazejewski's changing practice of veterinary medicine provides another peek into the quietly continuing urbanization of the American countryside along the highway, and the changing practice of animal medicine from the full range to the narrower speciality. In 1977, the office opened and spent all of its time treating the four-legged residents of the area's numerous horse farms. By 1982, when Stanley bought into the practice, horses comprised only two-thirds of the business. Today, horses make up zero percent of his work. The horse farms have moved west to cheaper land, hauling away the profits of their land sales to suburban housing developers tapping into the Philadelphia (twenty-five miles) and Wilmington (fifteen miles) markets. And in come onetime city residents whose blue collars are fading to white.

Besides their shiny cars, bicycles, and ubiquitous basketball backboards planted on the edge of their driveway, the new residents also bring their pets—no Chihuahua or other apartment dogs, no pit bulls to eat their youngsters. Stanley calls his a "companion animal practice" consisting of 50 percent cat work, 40 percent dogs, and 10 percent furry or feathery creatures like hamsters and parakeets. With a two-acre minimum typical zoning, the dogs are larger—Labradors and retrievers—and the care, he finds, is more meticulous than in the old rural countryside, where animals were tools and often had to fend for themselves.

The pets of Dr. Blazejewski's clients are members of the family, who receive the finest medical care, much of it preventative, that the human family members can afford. An entire family is likely to accompany a sick pet to the vet's; the waiting room is typically full of such family clusters orbiting around an ailing animal in an adult's arms. Amid the strange smells and sounds, each human offers his or her own brand of comfort to the creature and the adult pays with a credit card. The doctor must have in-patient visiting hours, as any hospital does, and a part-time groomer, although help is harder to find now, since where there is suburbia, there are teens, and where there are teens, there are Burger Kings paying six bucks an hour.

Nearing Maryland and still heading more west than south, stagecoaches hauling unwashed passengers and even Whitman's chocolates passed through one of the most historically perilous stretches of highway in the East, thanks to battling mercenaries and thieving highwaymen. At Chadds Ford (not a car dealer, this was where John Chadd and family carried travelers across the treacherous stream), Washington halted the retreat of his 12,000 men to try and stop the advance of General Howe and his 18,000 British and Hessian troops. It didn't work. The Colonials were outflanked and started to retreat again. But the Battle of Brandywine Creek on September 11, 1777, did provide the area with a future flow of tourists to visit the battlefield park.

The British plundering hurt the Chadds's fording business for some time, though not as much as the later bridge did. But the family had diversified into innkeeping, which was a respected calling in those days, offering as it did scarce safe haven to one and all during the dangerous vagaries of contemporary travel. Innkeepers were so respected, in fact, that they could usually get themselves elected to at least a major's rank in the militia.

Lodging was a minor part of the inn business, and it wasn't until the 1830s or so that travelers might count on securing a bed to themselves. More important was the inn's socializing role as local gossip central in the bar and the restaurant, which struck some foreign visitors as quite notable. One Francis Hall told his diary that innkeepers "were apt to think . . . that the travelers rather receive, than confer a favour by being accommodated at their homes." The inn's furnishings were invariably the same: a map of the new United States, a map of the surrounding state, perhaps a clock, a copy of the Declaration of Independence, and a mirror, brush, and comb for travelers to make their toilet before meals.

The meals were served at set times, come and get it or do without. Diners sat democratically at the same table without even a nonsmoking end, and all the food, soup through dessert, was set out at once, to save time. Although the inns did not advertise drive-thru or fast food, many diners could be filled and gone in less than fifteen minutes.

Breakfast was usually the traveler's main meal—buckwheat cakes, beefsteak, and cold apple pie constituted a typical morning meal. But the speed and vulgar manners shocked some Europeans. "You are, however, pretty sure against the conversation of unpolished people," one prince assured his peers in a missive, "because the Americans are usually mute at table."

Another well-traveled gentleman added in a letter: "A meal in the United States and Canada is simply a feeding, and not in any degree a conversational meeting." Stage travelers paid more than locals for the same meal because, the innkeepers explained with superficially infallible logic, the kitchen had to make extra food, not knowing precisely how many travelers would appear for a meal. No doubt, locals phoned ahead with reservations.

To help solve that problem today, the Chadds Ford Inn (yes, the same one, though the founder left us some 230 years ago) takes reservations and serves several main dishes from Cajun Catfish ($14.95) and Italian-style Grilled Chicken Breast ($14.25) to Lobster and Shrimp with Pasta ($19.95) and the Mixed Game Grill—grilled buffalo sirloin, boneless quail, and venison cutlet with Alaskan huckleberry sauce ($22.95). The old structure (1703) seats fifty-four for luncheon, dinner, and Sunday brunch in an old country atmosphere with prints on the wall from another family that has lived in Chadds Ford for so long, the Wyeths—N.C., Andrew, Henriette, and Carolyn.

Drinks are available at the inn and so are spirits. "Yes, that's right," says Dawn Jackson, the manager. "We have ghosts here. I've seen a little girl in a gown trying to light a candle on the mantel. One time, the water faucet went on by itself. And one night—oh, this is great—one night a half dozen people at the bar said they didn't believe we had a ghost named Katie. So I said, 'All right, Katie. Prove to me you're here.' At that moment, the cash register blew a fuse and the drawer popped open. You should have seen their jaws drop."

But it used to be scarier than that. The stretch of highway south of the inn on either side of the Susquehanna River was notorious for waylays. At least once or twice a year, a coach laden with people, baggage, and wallets would come upon some trees fallen across the road (yet another reason for widened roads). As the driver and male passengers struggled to remove the obstacles, two or three armed men with blackened faces would run from the roadside brush. They would tie the travelers to trees and then, since highway traffic was not yet exactly bumper-to-bumper, they would take two or three hours to rifle the wallets and mail. Cash was usually plentiful given the absence of credit cards.

One 1818 highway robbery provided its three villains with so much money (over $90,000) from the mail that they didn't even bother checking the driver or passengers. Such an outrageous robbery produced calls for more armed guards on stages. Meanwhile, special couriers were sent up and down the road, alerting authorities and merchants to the stolen bank notes. Within days, the trio was caught. They were tried. Two were sentenced to death. As usual in those days, part of the punishment involved the condemned writing a detailed confession, which was then widely distributed as a warning of the wages of sin. On September 10, 1818, less than six months after the crime, the two men were hanged in Baltimore before a crowd of 15,000. But, obviously, it wasn't until the railroad came in that American society was able to totally wipe out armed robbery.

This explains why in these times Bobby Jianney could obviously feel comfortable late on a Saturday evening pulling a huge wad of bills from his jeans pocket to make change for a gasoline purchase. But, of course, the gas station where he works is just inside Maryland—south of the fabled Mason-Dixon line. On the northern side, the self-service gas station requires payment in advance, and, with no intended irony over the association of its name with other bandits in history, the town is called Nottingham.

Turn Left at Maryland, Go Straight to the Caribbean, You Can't Miss It

CHARLES MASON AND JEREMIAH Dixon were two British astronomers engaged in the 1760s to survey a 132-mile section inland from the Atlantic Coast to settle once and for all some nagging boundary disputes between Pennsylvania and Maryland. At one time, Maryland controlled the territory as far north as Philadelphia. The Mason-Dixon Line, however, became more widely known as the boundary between the northern free states and the southern slave states. And while Maryland did not officially side with the South in the Civil War, it did harbor considerable Confederate sympathies. It was no coincidence that John Wilkes Booth lit out through Maryland after assassinating President Lincoln (and no coincidence that his alleged coconspirators included some Maryland natives, among them Mary Surratt. She grew up on the land that is now Andrews Air Force Base, became the widow of an innkeeper and an acquaintance of Booth, and, in the hysteria that followed the shooting, was the first woman ever executed by the federal government).

Inns still thrive in the Maryland countryside. On one recent Saturday night on 1 north of Baltimore, twenty-five motorcycles appeared to be feeding in front of the Little Falls Tavern and Logan's Inn had a good crowd for its country-western music. The highway is four lanes there before it narrows to pass the Earle Theatre (TWO ADULT HITS DAILY!) and a pair of liquor stores open late for thirsty folk. At the corner of Erdman Avenue and Belair Road shortly before midnight, one woman, who is no longer thirsty, negotiates the intersection on her hands and knees.

Tom Walters' Used Cars is closed at this hour, of course, although no one could tell by the bright lights. Tom, a veteran car salesman, struck out on his own not too long ago despite worries over the economy. Even in bad years, Americans still buy 13 million new cars. Bad times are actually when sales pick up for used-car dealers, because, while Americans will not stop using and buying cars, they may hesitate for a year or two to lay out the huge sums now necessary for new vehicles. Tom says many of his customers have apparently also hesitated over making a bill payment; two-thirds of them have some kind of credit problem (and so do a fair number of their banks nowadays). But Tom handles this obstacle by requiring enough down payment to cover his investment in the car and finances the remaining payments himself. So, at worst, he breaks even.

American buyers seem more interested in name quality, he says, so he's always eager to buy any Mercedes-Benz at auctions. Mercedes is a name that draws sellers. In the beginning of the automobile age, Americans took what they got for car model names starting with simple alphabet letters, *T* and *A,* very democratic. Very quickly, however, as cars came to be an important part of a driver's image, speed or the appearance of speed and sleekness emerged with a name like the Arrow. Americans then got into animals, the wilder and faster the better—Wildcats, Cougars, Lynx. So important did these names become that American car manufacturers ran out and patented all the wild-animal names they could think of, not the obvious losers, mind you (calling a new car the Possum would likely be the next Edsel of autodom), and not the Tiger (he was already working for Exxon and Kellogg's) but the sexy animals.

This was normal procedure in the United States. But it created a real problem when Japanese car makers began exporting big-time. It was okay for the Japanese back home to call some sleek model the Cherry. It sounded English to the Japanese ear and therefore carried considerable cachet, as French and Spanish can do to the American ear for cars (Coupe de Ville, Seville) and restaurants (Chez Louis or even Casa de la Maison but never Chez Burt). However, the American advisers to the Japanese car makers suggested that perhaps a name other than Cherry would work better over here. The Japanese accepted that advice and, with all the nifty wild-animal names already patented though likely never to be used, they had little choice but to make up their own exciting names—Tercel, Camry, Civic—turn them into nouns, and act as if the words meant something. Subaru is still having problems, having named one of its sporty little vehicles the Brat, incomprehensibly intending to associate itself with a spoiled child but, in the Midwest at least, instead naming its tiny vehicle after a German sausage.

My goal in life once was to own a Spider or just to sit in a Corvette, whichever came first. I ended up in a Valiant, which did not rumble loudly enough but did have a nice sound to its name and was red. I inherited this aging family vehicle for my nineteenth birthday. Alone in a rented room in a faraway city, I opened the parcel to discover a toy red car. How nice, I thought, my parents sent me a toy car for my nineteenth birthday. (In those days, toys came from Japan, not China). It wasn't until, as instructed by my parents' card, I went to insert the batteries that I discovered the Valiant's official title safely tucked inside. This caused great excitement in my room

and also taught me never ever even to consider throwing away a seemingly useless birthday gift. My teen best friend always wanted—and did get—an Impala, although any resemblance to that sprite wild creature by that big-assed Chevrolet was coincidental. No offense to the Japanese or my freshman math teacher, but personally, I could never present my automotive identity to the world in one of these new Japanese models whose name resembles an algebraic formula.

In Japan, they like black cars best, and brown. Americans liked black, too, because that was the only color cars came in until the invention of colored car lacquers in 1924. Nationally, the most popular color for sports cars remains black with white and red in second and third spot. But differences emerge in regional surveys. Sunnier places like Miami and Los Angeles prefer white the most, then other pastels, not least because the lighter colors reflect more heat (I think a black car in Miami in August might melt). Northeast sports car lovers like black the best. Every region thinks the fanciest luxury cars should be white (second place in the East and Midwest goes to black, in the South and West, silver). Nationally, pickup owners like any shade of red best, which I would have chosen, except they gave me such a good deal on the silver one.

Not far from Tom Walters's car lot, U.S. 1 loses its urban identity again for a few miles as it ducks and weaves through downtown Baltimore where, Southern-style, the houses and old apartment buildings are so hard by the narrow sidewalks, they barely leave room for two steps at the threshold. Baltimore was once the second-largest city in the United States not because of its highways or its impressive downtown urban renewals that, like Newburyport's, have returned the waterfront to the citizens. Nor because it was the hometown of Babe Ruth, H. L. Mencken, and Russell Baker (and Edgar Allan Poe's burial site). Baltimore was important because of its natural deepwater harbor on Chesapeake Bay and the waterpower from so many falls so close to town.

Long a major port (and still in the top half dozen), Baltimore was a hotbed of opposition to British attempts to regulate American trade, which have waned in more recent years but made it a major target in the War of 1812. But unlike its nearby sister city of Washington, D.C., Baltimore repelled the enemy in a battle off Fort McHenry, which, thanks to Mary Pickersgill, gave the United States its national anthem. Although Francis Scott Key wrote the anthem, he was inspired by the flag sewn by Mary, whose agent was less successful in garnering her historic credit than Betsy Ross's PR people back up 1 in Philadelphia.

Commerce, transportation, and communication have always been intense up and down the forty-or-so-mile corridor between Washington and Baltimore. In fact, the first telegram—"What hath God wrought"—traveled between Washington and Baltimore on Samuel Morse's new machine on May 24, 1844. But in recent times, the corridor has become so thickly settled and its interests, economies, and even professional sports allegiances so intertwined that the open countryside between the two cities is nearly gone. It is turning the corridor into another of the overlapping metropolitan areas that spread nearly the entire length of the Eastern Seaboard. Among census buffs, it is called Baltington.

The corridor's development began very early in the nation's history, simply as the shortest distance between the North and the South. The first private turnpike was chartered between Baltimore and the District of Columbia in 1796, just five years after the capital settled there. The Maryland legislature was conscientious in setting tolls for every conceivable kind of conveyance, including horses, asses, mules, and even camels, which was not to earn much money but certainly did add to the air of comprehensive coverage. Unfortunately, that first turnpike company did not get around to building a turnpike. So references in old diaries and newspapers are rife with complaints over the road's condition, from the bumps and mires of Colonial days to the thirties when a plague

of roadside signs prompted one team of researchers to title their report "The Disgraceful Approach to Our National Capital." (As if there's only one.)

In 1899, a special study by the Geological Survey had a good idea: "With good roads, mechanical motors of all kinds would come into much more general use and be a great economic advantage; and, moreover, the cost of maintenance of the roads themselves would be diminished as horses and heavy wagons were to a greater and greater extent replaced by automobiles with pneumatic tires." The survey's study also predicted that these new horseless carriages "would add greatly to the comfort of country life and would encourage many persons to live in the country who now crowd into the cities." So much for that solution.

U.S. 1 from Baltimore to Washington was zoned commercial in 1931. Ten years later, a survey found signs there running just under one hundred per mile. With the advent of electric signs and commercial strips, this visual static has not lost one colored watt. It passes through Waterloo and the former site of Spurrier's Tavern (GEORGE WASHINGTON'S HORSE DIED HERE), past the University of Maryland, a bevy of pizza parlors, hamburger stands, and bars, and old Bladensburg, where on August 24, 1814, 5,500 United States militia and marines failed to hold off 5,000 British regulars, who went on to burn the White House, the Capitol, and the State and Treasury departments. Bladensburg, once the flourishing head of navigation on the Anacostia River, now little more than a stream, was also the site of a flourishing early-nineteenth-century Dueling Ground. There, it is said, fifty-two duels were fought over some matters of honor that no doubt seemed important on those fifty-two dawns.

The dueling ground on 1 is in Prince Georges County, the now heavily suburban area of Maryland which donated a piece of itself to form the nation's capital, and so, too, is a developing chapter of the future, the little-known National Agricultural Research Center. Over the years since its founding in 1910, when the United States still considered itself a rural nation, the center has led vital research that has affected the lives of virtually every American, and millions more abroad. The 7,200-acre center, which does have weekday visitor tours, is deep into genetic and chemical engineering helping to fight crop diseases, boost crop yields, and pioneer medical research for both livestock and humans.

Scientists there have developed different methods of planting that fight soil erosion. They have produced natural insecticides, viruses that kill voracious pests destroying vast amounts of crops, and all without damaging chemicals. They have isolated chemicals given off by female insects when ready for mating, as well as bits of brain chemicals that control an insect's maturing, which contributed to developing effective new bug traps or tricking them out of the mating mood. They are improving breeds of animals, including one kind of cow that may someday give only skim milk; and hogs, whose hair is used for brushes, whose hide helps graft skin onto burn victims, whose pancreases produce insulin for diabetics, and whose heart valves are used in human heart operations.

Some of the experiments have aroused the ire of the increasingly militant animal-rights activists, who broke into one lab a few years ago to free some creatures. But the center is also developing dwarf fruit trees that produce greater yields per acre with less harvesting cost than larger trees. They are working on strawberries that bear fruit in the fall, to spread that season out for farmers and consumers. They are developing methods of combating fire ants, those industrious little critters who arrived in Mobile, Alabama, around the turn of the century and have been working their way up through the Carolinas ever since, building nesting mound after nesting mound, ruining farm machinery, attacking and killing infant livestock, and undermining countless roads with their intricate caverns.

On one two-acre test plot, agronomists are even developing new grass mixes that offer hope for cities where people do walk on the grass. The test involves mixing tough zoysia grass, which can stand summer heat, drought, and lots of shoe soles but turns brown at the first frost, with tall fescues that sleep during the heat and green up in cooler temperatures. The idea is to have a tough year-round green lawn so packed with luxuriant slow-growing grasses that weeds can't grow and mowers need not go often. It is being tested, among other places, on the Mall in Washington, D.C., where the shuffling shoes of one night's band concert can wipe out 200,000 square feet of costly sod.

Washington is an interesting place if you don't live there. In its most recent itinerant incarnation, U.S. 1 enters the city as a broad Rhode Island Avenue not far from the dueling ground on the northeast side and wanders through a neighborhood of aging one- and two-family homes and small groceries before veering down Ninth Street behind Ford's Theater, where President Lincoln was shot, and alongside the FBI headquarters across Pennsylvania Avenue to Constitution Avenue at the Mall. At Fourteenth Street, 1 turns south again through some minor marble canyons by the U.S. Mint, which prints $30 million a day, plus postage stamps for the United States and many other countries. It darts (at non-rush-hour times, at least) past the Jefferson Memorial and across the Fourteenth Street Bridge (where the passenger plane crashed a few winters ago) to march across the Potomac River into Virginia.

According to a plaque that once adorned the roadside on the Maryland boundary, George Washington first entered Prince Georges County, as then constituted, in August of 1751, and "made his last exit therefrom December 18, 1798. Ave, Ave, atque Vale: Hail, Hail, and Farewell." In the tradition of aging chief executives contemplating retirement, Washington chose a compromise capital relocation site that just happened to be an easy ride from his country home. Washington named the capital Federal City. Congress, perhaps worried that that name sounded too much like a bank, waited until the first president was dead and then changed the name. Washington (WASHINGTON DID NOT SLEEP HERE) wasn't much of a city even then, but the name change did inaugurate another hallowed American tradition of naming places for dead people.

Federal City was designed by Pierre L'Enfant, a French engineer on General Washington's staff, who did such a good job of laying the beautiful place out in broad commonsense spokes that Congress inaugurated another tradition to celebrate the fruits of rational thought; it fired him. Over a century later, L'Enfant was honored for his pioneering work drawing up the first city ever designed for a specific purpose. But by then, he couldn't make the ceremony.

The District of Columbia was carved from one hundred square miles of Maryland and Virginia; there were only 130 federal employees in those days, but they saw the potential. In 1846, Virginia got its thirty-one square miles back for future suburban development. Although Mr. Washington laid the cornerstone of the Capitol in 1793, the government remained in Philadelphia until 1800. Then, during the administration of President John Adams, who liked the place, but amid considerable grumbling from others about abandoning the comforts of eastern Pennsylvania for a crude frontier fort in a swamp, the government packed up its file cabinets and feather pens and moved down U.S. 1 to its present site. Fourteen years later, the British came, the British came. But Washington was rebuilt. And although for a long while it appeared to contain an incongruous amount of marble for a standard swamp, Washington eventually evolved into what it is today, the nation's capital of politics and homicide, where southern efficiency meets northern charm.

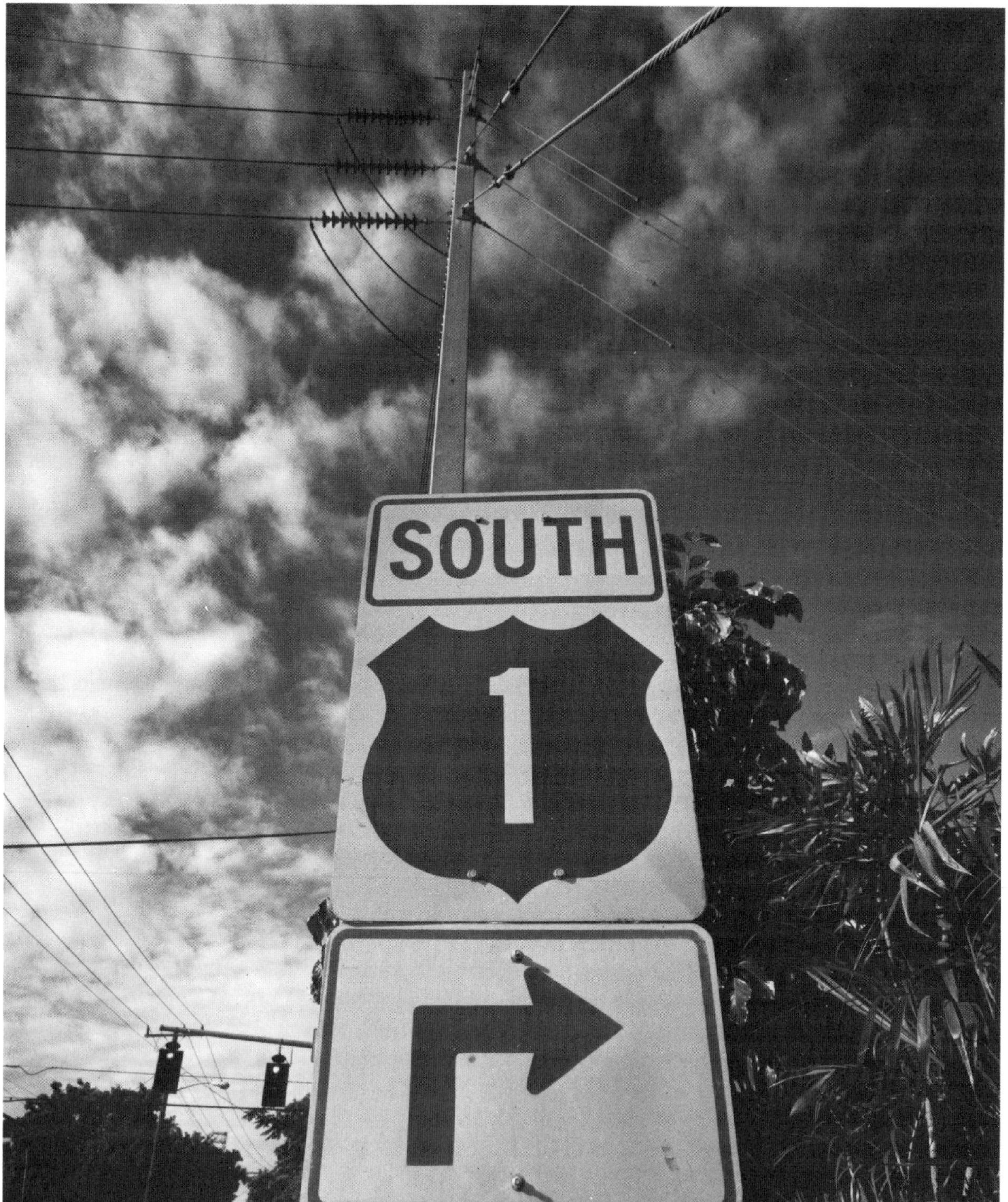
SOUTH
1

PART III

VIRGINIA TO FLORIDA

For God and Country and the Future of Tourism

HOW MANY VIRGINIANS DOES it take to change a light bulb? The answer, of course, is four—one to change the bulb and three to talk about how great the old one was.

An awareness of history, and its economic potential for the future, is indeed a characteristic of the Old Dominion, where the first permanent North American white settlement was established at Jamestown in 1607, according to Virginia tourism authorities, who did not submit their claim for approval by St. Augustine tourism authorities. Virginia positively reeks of history, much of it right along the shoulders of U.S. 1, much of it still celebrated today, and much of it noted on the 1,500 historical markers that line Virginia's roadsides. The state has produced eight presidents, not to mention Patrick

Henry (or Robert E. Lee), the authors of the Declaration of Independence (Thomas Jefferson), the Constitution (James Madison), and the Bill of Rights (George Mason). Virginia was the scene of the first Thanksgiving, the first armed rebellion against the monarchy (1676), and fully half of the fighting in the United States's first civil war. Some meaningless and many meaningful military campaigns were fought on Virginia soil beginning with the Revolutionary showdown at Yorktown in 1781, and running through the bloody Civil War, which remains the worst fighting in North American history. At the famous battles of Fredericksburg, Chancellorsville, Spotsylvania, Bull Run, and the deciding struggle at Petersburg, so many of the boys who had traveled all the way down from

Fort Kent and Caribou finally met their Maker, although so little remained of their remains that identification in most cases was impossible.

The historic highway that runs nearly two hundred miles through Virginia's Tidewater region was first the proverbial Indian trail, which the whites called the Potomac Path (Pocahantas was taken along it), and then it was a highway for troops on both sides of the Civil War. Washington rode on it, and Monroe, who kept his law offices in Fredericksburg, and Jefferson, who like Franklin was very busy inventing things all over but took the time one day to feather a complaint about the troubled road: "We could never go more than three miles an hour." That's *twice* the speed of crosstown traffic today in New York City, and slower even than the average five-mile-an-hour pace of the Northern and Southern armies as they made their way to Washington after the truce at Appomattox. Office-working commuters heading into the Washington area from Virginia on the same road today would be delighted to break that speed barrier during some rush hours along the four-lane, sometimes divided, always surburban U.S. 1. The road enters Virginia from Washington on the off-ramp of the Fourteenth Street Bridge not far below Arlington National Cemetery and right next to the ominously gray headquarters of America's military, the Pentagon.

With flowers perfectly arranged in the median, the highway moves through a newer corridor of hotels and apartments called Crystal City, beneath the studio window of the nationwide Larry King radio show, for a shiny mile or two before plunging into the kind of dismal old railroad-yard neighborhood where the Department of Motor Vehicles in virtually every state seems to set its driver testing stations (one of the requirements established by bureaucratic regulation is that the doors to these stations do not open to the public until the line is thirty deep or the clock says 9:30, whichever comes last). The first stop for the famous road, which will have many names, including the Jefferson Davis Highway, before it

just disappears in Key West, is Alexandria, another one of George Washington's hometowns.

Alexandria was formally laid out as a seaport in 1749 by Scottish tobacco merchants. John West, Jr., a surveyor, with his young assistant, George Washington, had planned the streets and half-acre residential lots. And in tribute each February, Alexandria holds a month-long birthday party with what it believes to be the largest George Washington parade in the country.

The highway out of Alexandria is standard suburbia strip with muffler shops, dog kennels, gas stations, movies, and auto-body shops mingling with no-name motels, costume-rental stores, occasional outbreaks of palmists, and enough 7-Elevens to serve a population twice the surrounding size. There is, according to one corporate study, one chain restaurant for every 2,250 Americans. The people of Dallas's Southland Corporation, 7-Eleven's corporate parent (yes, another of those popular leveraged buyouts), say they do no not know how many of their 7-Eleven convenience stores are located on U.S. 1. However, a rough personal estimate after many long, tiring days is that of North America's 6,632 7-Elevens, it certainly seems that at least 6,027 are placed along U.S. 1. At some points in suburban Virginia, 7-Elevens emerge like dandelions at the rate of one every 1.7 miles. In fact, after California, Virginia has the most 7-Eleven stores of any state (674). Third place goes to Florida, not surprising given how important warm weather, leisure time, and heavily populated urban areas are to those stores' mix of three thousand products. There are, for instance, no 7-Elevens in Maine at the top of U.S. 1, while at the bottom, Florida has 621.

The chain and what it offers the 7 million people who walk in the door every day reveals something of American priorities these days. First, the store must be on a main street with a lot of parking; E-Z In-Out to match the Interstates' E-Z Off-On. Even though less than half its stores sell gas, 7-Eleven still pumps about 1.5 billion gallons a year, nearly 20 percent of its total sales.

(Southland used to own Citgo, too, but sold its interests to Venezuela's national oil company.) Second, the stores' typical twenty-four-hour operation must offer a variety of quick sale items; tobacco products make up almost another fifth of sales. Drinks—soft and hard—compose another 21 percent. The company is trying to broaden its offerings—adding more fast-foot items such as pizza, hot dogs, deli-style sandwiches, and one of the all-time most profitable food items, popcorn. Automatic bank tellers have been installed to make it very convenient to get money to spend in the convenience store, along with rental movies and electronic games, magazines and newspapers, of course, and even postage stamps. The chain has become the undisputed top sales outlet nationally for state lottery tickets. (Technicians are close to perfecting the latest breakthrough in the vending-machine industry that began with tutti-frutti gum in 1886; if we as a people can just hang on a little longer, we will have the french-fry vending machine.) All of which adds up to give 7-Eleven, this modern franchise equivalent of the general store, about $8.4 billion in annual sales, making it one of the nation's largest retailers. Not too bad for an operation that began by selling a single prized commodity in Texas—ice.

Convenience, however, can have its drawbacks. After buying their Slurpees, many youths tend to hang around outside, ominous groupings that can prompt some former teenagers to drive right on by. To combat the growing problem of teen loiterers, McDonald's eliminates one major teen attraction by purposely not installing pay telephones. Such a created inconvenience, however, would seem rather counterproductive for a convenience store. So some 7-Elevens have come up with an effective alternate strategy: They use outdoor speakers to play loud elevator music. Young people make their purchase and then flee in the face of music that does not require electric guitars and drums.

Interspersed in this road's corridor clutter is the occasional woodland or a surviving piece of history. In Virginia, near Accotink, on the left side, is the tiny Pohick Church, where Washington maintained two pews. According to the custom of the day, the wooden pews had high backs to protect the worshiper's privacy but it also avoided publicly embarrassing those who might nod off (WASHINGTON SLEPT HERE).

Roads have always been a powerful economic attraction; cities have grown up where dirt paths once crossed. Virginia's were not too good when the proper ladies of Washington rode their carriages into the countryside of an afternoon to view Civil War battles arrayed for their pleasure. And the mud-choked lanes were not too good even seventy-five years ago when the Automobile Club of America urged tourists to avoid the state because its roads were so bad. Even when the state did fix up one part of old 1, some local citizens came out at night to destroy the improvements; it threatened their thriving tow-truck businesses.

Back then, new and better roads were touted as panaceas that would break down regional isolation, enable farmers to get their produce to the markets of crowded cities more quickly and, thus, keep more farmers down on the farm (another governmental solution that worked really well). Road building was an economic spigot that government could turn on and off to spur growth and combat decline. When the final link in U.S. 1 from Washington, D.C., to North Carolina through Virginia was dedicated on November 25, 1927, it was so important that both governors showed up with their wives and the Richmond *Times-Dispatch* devoted most of its front page to the red, white, and blue event; inside, the stories took up most of a page except for the sale at Thalhimer's on Women's Linen Knickers—$2.95 each, sizes 16 to 20, "Suitable for all sports."

The governors differed in their approach to financing roads: North Carolina liked bond issues: Virginia favored a more conservative idea, called "pay as you go." But both agreed on the road's immense value. Governor Angus W. McLean of North Carolina credited advances in his state's education system to parallel improvements in its highways. And Governor Harry F. Byrd of Virginia promised further improvements to the world's heaviest-traveled road, because it would help bring more "desirable" residents to his state. Governor Byrd then took Governor McLean to view the road from an army blimp; it was the North Carolina governor's first airplane ride.

At first, merely traveling on such roads was an end in itself. Automobilists would daringly duck out into the countryside for a day, possibly overnight, but always in a group for protection against breakdowns and whatever else they thought lurked outside the cities. The joy of moving freely through the fresh air in an open car, liberated from the smells and fanny-jamming jars of a moving animal's bony back, seem in old writings somewhat akin to the excitement and liberation now felt from such potentially daring and open-air activities as water-skiing, downhill skiing, and hang gliding. I still feel that way on warm Saturday mornings.

Even my dog Buddy shared that exhilaration in our mutual childhoods. He loved to sit in the backseat of our car in the driveway. If anybody opened a car door, Buddy, the nongenteel St. Bernard, was right in there, perching his bulky body pertly in the back, ready to go. Even when he wasn't invited, Buddy would suggest getting in the car, as when my father arrived home from work, tired and wrinkled, set the car's brake, removed his keys, and turned to confront a huge dog trying to climb through the driver's window. Buddy had very strong second thoughts as soon as the vehicle actually began to move. But once we were cruising down some highway, he, like me, had his head out the back window, hair flying in the breeze, eyes squinted, mouth smiling. He might even bark. And I would yell, so exhilarating was the sense of effortless movement.

As a child, the most important thing was to be moving—"I move, therefore, I am." My grandmother called it the fidgets; you caught them like the flu. By my teen years, of course, it was more important to be seen to be moving. And it had to be moving on your own, none of this parent-driving stuff. This independence was accomplished, naturally, by getting a driver's license, a crucial rite of passage which the state thinks has something to do with certifying driving skills. It is, however, more like going on your first hunt in prehistoric days; you are on your own, accepted, mobile, with the license, keys, and wheels to prove it. My dream once was to be seen to be moving—probably haughtily and certainly with the engine rumbling deeply—down main street in a '56 Merc with the correct radio station blaring louder than parents would like. But, *ah hah,* they can't do anything about it because I have my license now, so I can be wherever I want, which is wherever they aren't, almost anytime I want, which is heady stuff.

Historian Daniel Boorstin has called the automobile the "anywhere vehicle," and it is ideally suited for the age of the everywhere community. Go here. Go there. When you want. How you want. Come back sometime. Anytime. In a car, lines of class, education, occupation, even gender are erased. You are what you drive. To be sure, one segment of society remains ostentatiously disdainful of mechanized mobility. But the United States is the most automobiled society in the world, 1.8 people per car, compared with 2.2 in Canada and 96 folks per car in Thailand, which would make for a mighty big car. Besides going to and from work, the likeliest reason for car travel is to visit friends or family; not surprisingly, the busiest day for that activity is Sunday. Interestingly, however, the second-busiest day for visiting by car is Monday, an indication of the growing popularity of longer weekends. This is a phenomenon not overlooked by the hotel industry, which has traditionally focused on the regular business traveler and, thus, found itself largely empty on weekends. So watch for more emphasis on weekend getaway packages as the powerful travel industry seeks to mine the profits of two more statistics: Americans are taking many more trips of shorter duration, and while vacations make up less than 2 percent of total trips, they comprise nearly 20 percent of all mileage.

Nearly 80 percent of all American travel is still by

car. In the average year, about 166 million licensed drivers in 89 million households make a total of nearly 130 billion trips in 192 million vehicles. Each vehicle averages about 11,000 miles, which means that every year Americans drive a combined distance equal to 310 trips from earth to the edge of the planetary system *and back*.

Every year between 40,000 and 50,000 of these travelers don't make it home, half of them due to alcohol. That's a total of *2.7 million* fatalities during the age of autodom. Total damage to people and property paid out by the country's insurance companies, including those back up 1 in Connecticut and Rhode Island, is more than $24 billion a year. The safest place to be on American roads is in a full-size car on an Interstate at 8:00 A.M. on a January Tuesday. The most dangerous place to be is on a motorcycle anytime anywhere. But among four-wheeled vehicles, the least healthy place is on a two-lane local road in a compact passenger car on a July Saturday night at 1:00 A.M. (although the wee hours of January 1 are no picnic, either).

But despite such highway carnage, American roads, thanks to Interstates and expressways, are the safest in the world by miles driven. By population, Syria is the safest driving country. But Syria's figures do not include deaths from car bombings or passing machine gunners and are probably further skewed by the notorious inaccuracy of odometers in maneuvering tanks.

So popular has motoring become that what began in 1902 as a lonely little grouping of nine auto clubs with fewer than one thousand members among them has become one of the largest single membership organizations in North America, the American Automobile Association. And so large has the AAA become with its more than 32 million members that it had to leave its old headquarters just off Route 1 in Virginia to find more spacious and more affordable space just off the same highway down in the Sunbelt.

With 35,000 employees in this nonprofit federation of 150 affiliated motor clubs, the triple-A claims as members 15 percent of the adult driving-age population who own one of every five cars in the United States. It writes more than $3 billion in automobile insurance, produces more than 320 million travel-related publications annually, is the largest seller of American Express travelers checks ($2.5 billion a year), oversees nearly 14,000 road-service facilities, and sponsors numerous safety programs, including 500,000 school safety-patrol members. So much a part of their life has the AAA become for members that the automobile club sells more airline tickets every year than any other travel agency, except the airlines themselves.

After a four-year study of eighty-three metropolitan locations, the association made a decision that many companies have made in recent years: it moved south down the highway for the lifestyle, the reduced urban hassle, and the real estate values that for the same price let it buy half again as much office floor space outside Orlando, Florida, as it had in suburban Washington, D.C.

Americans need to belong to things and to celebrate that membership, whether it's an allegiance to a professional sports team, an extended family, or claiming a blood connection to a lost religious zealot who happened to have stepped on a particular rock near Plymouth. The AAA has been among the most successful in attracting members, wherever it is—a million new members each year, about one every two minutes. "The pendulum is swinging back to traditional American values," theorizes Richard Hebert, a managing director. "Americans are returning to some basic values in the family, in the foods, in many things. And auto travel is like an oatmeal cookie, a solid American favorite, not some faddish chocolate-covered truffle."

But somehow as the years have passed and our ability to collect and store and collate all of this and even travel at the speed of sound near the troposphere, the vivid joy of simply doing in our society often seems to become more the pale satisfaction of having done—or watching someone else do. Racking up numbers, not experiences. Seeing six states in four days and having the photos and souvenir window stickers to prove it, but not many memories to cherish beyond the paid-admission gate of pre-packaged theme parks with their ersatz exotica.

Now, we frequently even count the air miles because we can get free tickets simply by sitting in the full upright position (with tray tables safely stowed) seven miles above the nearest visible highway. Up there, there's no dispute about wearing the seat belt that, if we follow instructions to pull firmly on the tab, will hold our bodies safely within the foam-rubber confines of a cushion that turns to cyanide gas when exposed to flames. But, on the other hand, we can see nothing as our aluminum cocoon covers one mile every six seconds. We do get airline meals, the exercise of running to a connecting gate, and the chance to share a plastic armrest with someone else's talkative relatives. No need for roadside picnic tables anymore, unless there's a phone there so we can get something else done, too, during our rest stop. Thank goodness, they finally got around to putting telephones in airplanes, lest there be some moments someplace free of telephonic touching.

We don't seem to treasure the road miles so much anymore, only how quickly we can get through them to the destination. "Gee, I don't know how far it is, but it'll take you about an hour—depending on traffic, of course." In fact, of the average year's 8 million requests for AAA highway-routing suggestions, today barely 16 percent ask for the scenic way; 84 percent want the fastest direct route possible—which isn't all that fast anymore given all the reconstruction in progress. But it looks direct and orange on the map and combats the dread disease—dawdling. And, like as not, upon arrival we now take it for granted that our destination doesn't look much different from our departure point. Why should it, right?

Unless you live in a secret test-market city, the menus for McDonald's, Burger King, Hardee's, Roy Rogers, Arby's, Pizza Hut, Red Lobster, and Denny's are the same familiar fare. The predictable is profitable. Who would ever want a hamburger that tasted differently in Rhode Island from one in Georgia? Without Duncan Hines, what traveling stranger would risk a Moody's? We can sense which Interstate interchanges will have what

we want. And I don't know about you, but I can get around in a darkened standardized Holiday Inn room about as well as I can in my own home, which does not have a toilet seat wrapped in a thin paper strip certifying the plastic oval has been sanitized.

In today's travel industry—one of the nation's (and Virginia's) largest businesses right up there with education and defense—the idea seems to be to straighten the road, eliminate all surprises, which is great for business travelers, a majority of whom use cars on their business trips. The other goal is to make everywhere a destination. That's the way to get the most tourists—and the most money, which go together.

"Tourists will come to Virginia and we must not only give them roads to our historic shrines, but we must mark these shrines, plainly, briefly, and accurately," Governor Byrd said at that 1927 highway dedication. Official projections of American road use are almost always wrong; when the Pennsylvania Turnpike opened on October 1, 1940, they predicted 1.3 million vehicles would try it that first year. They got 2.4 million.

Fredericksburg has sought such tourists with another kind of dedication. A one-time fort, musket-making center, and tobacco port until the ships got too large for the Rappahannock River, Fredericksburg has become home to a burgeoning tourist industry (and a major 7-Eleven distribution warehouse). So important are cars to tourism that the mayor of Fredericksburg signs and issues free downtown parking passes to any and all of the community's 250,000 annual visitors. That's 2.5 times more people coming to see a battlefield than were killed or wounded there in the first place. (And so important are newspapers in igniting these tourist invasions that the city of 15,000 has its own Tourism Department, which provides, free of charge, pictures for publication of virtually anything in the city—in black and white eight-by-tens or thirty-five-millimeter slides, for vertical or horizontal layouts.)

Some outlying parts of Fredericksburg don't look much different from any other old community. Nearby are the usual arrays of fast-food outlets, carpet cities, used-car lots, and gas stations, none of them alluding to the famous little two-act drama played out near there. Pocahontas, daughter of a chief, was kidnapped by Captain Argall from an Indian village in 1613. The plan, which was not considered terrorism then as long as the kidnapper was white and the kidnappee was something else, was to hold the princess for ransom. But instead, she fell in love with one of her captors, which ruined everything except the fable.

The aging strip malls along Route 1 require renovation. But the grassy roadside fields are the Saturday scene of countless parents working patiently with bands of tykes in soccer shorts that were white when their mothers laid them out that morning. Like water bugs, the youngsters run every which way in clusters of chaos while the parent coaches yell instructions to no attentive ears.

But downtown, Fredericksburg reeks of a well-kept, revered, and profitable history, with Ye Olde signs, Ye Olde stores and furnishings, and Ye Olde costumes, even on the tavern wenches and the volunteers manning and womanning Ye Olde Visitors Center with its brief un-Ye Olde "Historic Fredericksburg" video perpetually projected in the viewing room. The friendly center has walking-tour booklets for virtually every street and booklets for cyclists. Outside, a belaced ship captain in tight knee pants, large-buckled shoes, and brass-buttoned coat swaggers along the sidewalk, greeting the gentlemen, tipping his tricornered hat to the ladies, and tousling the hair of visiting youngsters.

He's really Tom Lindsay, a retired cross-country truck driver who left the highway in 1979 to become his true self, one Captain Yorkshire. At the age of thirteen, Yorkshire was shanghaied and became a cabin boy and eventually a captain. The good captain gives summer walking tours to one and all, strolling the old pavement with energy and waving his lacy cuff at the passing attractions. There is the law office of James Monroe, with the famous Monroe Doctrine desk, immortalized since 1823, and a copy of that document's famous but murky prose. There is the Apothecary of Hugh Mercer, doctor to the Revolution, and the Rising Sun Tavern of Charles Washington, George's brother, who had to settle for having a West Virginia city named for him. And, of course, there is Kenmore, the home of Betty Washington Lewis (WASHINGTON'S SISTER SLEPT HERE). Betty only got a line of Sears appliances named for her house. But she looks more like her puffy-cheeked brother than he does, or did. In fact, the similarity is so powerful that George is said to have thrown his cloak over Betty's shoulders, plopped his Revolutionary hat on her short brown hair, and vowed to send her off to fight the British. (If a TV crew had caught that endearingly human scene, George might still be president.)

Mary Washington, the mother of the father of our country, also lived in town and has her own spire-shaped Washington Monument. George chopped *the* cherry tree in Fredericksburg, although current environmentally correct lore makes more of his planting chestnuts; green was, in fact, his favorite color. The Reverend James Mayre tutored little George there (WASHINGTON DID HIS NUMBERS HERE) in the days before schoolchildren used calculators in third grade. All summer long, Captain Yorkshire tells these tales to endless carloads of tourists: how Fredericksburg was a frontier river fort right by the falls, how George grew up there before moving to take over Mount Vernon upon the death there of his older brother Lawrence. (Do you suppose that when they weren't posing for portraits and speaking for history books, these men of the first First Family called each other Pudgy, Larry, and Chuck?)

Captain Yorkshire tells how Lafayette walked those same sidewalks and how it wasn't a dollar that George threw and it wasn't a northern river he lobbed it across; it was a Spanish coin and the river was actually the Rappahannock just over there. No one explains such profligacy, however, and no one makes much of who the original Frederick was—father of the notorious George III. Come winter, Captain Yorkshire changes costumes to become General Washington.

Halfway between the Union capital of Washington and the Confederate capital of Richmond, which is on a major railroad, lies Fredericksburg. Its most dramatic role in history, however, was being the site of four major Civil War battles—Fredericksburg, Wilderness, Spotsylvania, and Chancellorsville, when Southern soldiers hurt their cause so badly by fatally shooting their own Stonewall Jackson. In the Battle of Fredericksburg, Union General Ambrose Burnside gave General Lee his most one-sided victory in a five-day affair centered on December 13, 1862.

The Federals had marched on the town of five thousand as part of their blind "Take Richmond" strategy. President Lincoln understood the psychological import of seizing the Confederate capital. But he was a veteran of fighting Indians on the western frontier, where the enemy's capital was a portable collection of tents. So the president kept suggesting that perhaps the most important goal should be destroying the Southern forces wherever they were found. This strategy of massive destruction fit the approach of two Union officers, Ulysses S. Grant and William Tecumseh Sherman, who would have their chance—and their victories—later farther down this same highway. In fact, when Grant twice encountered military stalemates eighteen months later near Fredericksburg, at the Wilderness battlefield and Spotsylvania Court House, he did something unthinkable: He went around the enemy.

But General Burnside had decided on a frontal assault, and so a frontal assault was what it would be. He could have waded across the river on his arrival. But he waited for his portable bridging materials, which were late. This gave Lee's men time to rush up and dig in. Then the rains came and the river rose. And the fates were sealed for thousands of men who were not back in the command tent.

More than 15,000 men are buried at the Fredericksburg National Cemetery, nearly as many as live in town now. The identities of more than 3,200 soldiers will never be known. Hundreds of other bodies were never found. Battlefield medicine for the wounded was crude indeed in those days, though dedicated care was provided by orderlies and volunteers such as Clara Barton, who founded the American Red Cross on the bloody Fredericksburg battleground. Many casualties fell at Marye's Heights (John Marye's front porch was a Southern observation post). At the foot of the heights stood a shoulder-high stone fence. The Rebs stood up behind there and, volley after volley, blasted away at Union assaults. Seven times, the bluecoats charged across the open field into the Southern barrels. The Yanks, one Union soldier recalled, were blown back in a wave "as if the Doors of Hell had opened."

The fields were soon littered with the grotesque human debris of battle. Like many American units of the day, for instance, the Irish Brigade relied heavily for recruits on newly arrived immigrants, who had fled the turmoil and deprivation of Europe. When that brigade began the battle on December 13, it had 1,200 officers and men. The next morning 280 remained. Throughout history, noncombatants had only been able to imagine such scenes in colors sanitized by ignorance. But there was something called photography beginning to wander these battlefields. Its vivid legacy and that of its technological descendants would forever change the political and military course of combat, not to mention provide brisk sales at the nearby bookstand of the National Park Visitor Center. Across Lafayette Boulevard is the Battlefield Restaurant, the Stone Wall Souvenir Shop, and Kenny's Auto Brokers. Out back beneath the old trees, right where strangely silent tourists now park their cars and hush their exuberant children newly liberated from backseat seat belts, stands that very same stone wall, still shoulder-high.

The sure sign of creeping suburbanization erasing rural America are the yellow backhoes on the roadside, planting new sewer pipes in the old ditches that have provided adequate drainage for so long. But with so much more pavement going in for streets, driveways, foundations, and parking lots, there's less soil open to absorb the runoff. So the new pipes go in with the new housing tracts that erupt like boils behind a thin wedge of trees. Someday residents may want to screen themselves from the bustling of U.S. 1. But for now, the developers erect long series of large and small roadside placards proclaiming the virtues of their properties JUST AHEAD, like the Burma-Shave signs of yesteryear.

These modest signs, with sometimes corny, sometimes clever verses, provided pleasant chuckles to nearly fifty years of motorists at a time when cars went slowly enough to read the series of six or seven small red boards, always ending with the shaving cream's trademark logo.

WITH GLAMOUR GIRLS / YOU'LL NEVER CLICK / BEWHISKERED / LIKE A BOLSHEVIK / BURMA-SHAVE.

IF YOU / DON'T KNOW / WHOSE SIGNS / THESE ARE / YOU CAN'T HAVE / DRIVEN VERY FAR / BURMA-SHAVE.

The signs not only launched an instant shaving lather, they played a crucial business role in showing that humor could sell, and so could short, snappy jingles. So much a part of popular culture did the signs become that millions regularly ended every joke with the line "Burma-Shave." And this, generations before anyone said, "Where's the Beef?"

Burma-Shave began appearing in the mid-1920s in half-pound jars as the first brushless shaving cream. Invented by Clinton Odell after more than 140 different chemical tries, it eliminated laborious mixing by millions of sleepy men making their morning ablutions. Ignoring the advice of professional ad men, who favored large blocks of somber type, Clinton and his brother Leonard bought some used lumber, painted crude slogans on them, and pounded them into the shoulders of the highway just outside Minneapolis. Within weeks, orders were flowing in from many druggists who had one thing in common: Their stores were located in towns all along that same highway.

The signs caught on and spread and so did the shaving cream. Eventually, more than seven thousand sets of signs were standing, one hundred paces apart, along the roadside of almost every state. Alexander Woollcott once said it was as difficult to read one Burma-Shave sign as it was to eat one peanut.

Like the McDonald brothers' hamburgers, the signs and shaving cream were so successful, they attracted powerful outside interest. Guess who bought Burma-Shave? The Philip Morris conglomerate, the world's largest consumer-products company (Kraft, General Foods, Miller beer, Maxwell House, Marlboro), which decided to abandon the roadside ads for more modern mass means. This made one of the shaving cream's verses seem prophetic:

SHAVING BRUSHES / YOU'LL SOON SEE 'EM / ON THE SHELF / IN SOME MUSEUM / BURMA-SHAVE.

Today's roadside signs can produce more gasps than chuckles, like the billboard not far from Fredericksburg for Lyndon Larouche's political movement. Addressing the President of the United States, its eye-grabbing large type says simply, "Eat It, George." On one side is a picture of a piece of broccoli.

Maybe thirty miles before Richmond, the Park 'n 'Ride lots start, not far from the hometown of Patrick Henry, a hard-luck small-town lawyer who would turn a single phrase into eventual fame—and have a nearby motel named in his honor. Confederate flags begin making an appearance along with the first of many old—and I mean old—abandoned gas stations, the ones with the twin wooden pillars out front and the roof that covered the roadsters making their way along U.S. 1. Not far along, just down from the Green Top Gunshop, is where John Prather patrols.

John is the caretaker for the Evergreen Pet Cemetery, one of those less obvious signs of modern America's attachment to once-wild creatures as obedient companions. For $365, a grieving pet owner can get a twenty-four-inch casket, a burial, and one-year's plot care (perpetual care goes for another $300).

John oversees 1,500 graves, including one for a

horse. Amid the sounds of crickets and songbirds frequently drowned out by passing diesels, John paces the paths between graves most days, checking the grass, the stones, and removing the faded bouquets that adorn most plots.

There are a lot of Fluffys and Queenies and Skeeters resting there. Also Muffet, LOVED BY THE CASTENS 1971–1987. Fluffy Burgess, A SPECIAL FRIEND, lived to be almost nineteen. Sambo Jackson Bullock did not live that long: BORN DEC. 12, 1979 DIED MAY 8, 1980 MOM AND DAD LOVE YOU.

John says business is off considerably what with all the layoffs and shuttered factories around Richmond's old industrial sector. The state prison at Spring and Belvidere streets is busy, though. Richmond was first settled in 1607 and ever since has been the center of much of the life (and death) of Virginia, which once covered a vast area reaching as far west as the Mississippi River. The city was a much-fought-over territory as Indians fought with whites, Colonials fought with British and Germans, and Americans fought with Americans. St. John's Episcopal Church there was the site of what might well have been the Revolutionary War's best sound bite, Patrick Henry's "Give me liberty or give me death" comments. The former capital of the Confederacy, Richmond remains the capital of Virginia and is, by the way, the country's single-worst market for coffee, according to per capita statistics kept closely by the nation's brewers (New Haven and Portland, Maine, are among the best, in case you are keeping track, too). This may reflect absolutely nothing of significance, along with Richmond's proximity to an old stagecoach house on the road to Petersburg, which is locally famed as the birthplace of the mint julep.

Various retreating bands of Southern troops fired parts of Richmond, as did Union artillery. But it was an unsuccessful Union assault in the spring of 1864, the costly Battle of Cold Harbor, under the new chief military commander, U. S. Grant, that prompted the North to abandon its frontal attacks. While the North had more men to spend than the South, the costs to the blue were

far too great. So Grant set his sights, literally and figuratively, thirty miles to the south on an unassuming place named Petersburg.

During the Revolutionary War, Petersburg was pillaged by the British (that New Haven traitor Benedict Arnold again). Yet another former frontier fort, Petersburg had diversified beyond tobacco agriculture and commercial trading into the developing industries. Not only did it have tobacco warehouses and cotton and flour mills, it also had iron foundries. And local production had cheap access to world markets down the Appomattox and James rivers. As the Civil War approached, the city's locomotive works produced hundreds of cars and engines for the *five* different railroads that met there to send a single line into Richmond. "The key to taking Richmond," Grant once wrote, "is Petersburg."

While appearing to continue his long-distance stare at Lee near Richmond, Grant quietly moved groups of troops toward Petersburg. On June 15, he sent a superior Union force into the undermanned Southern lines. They were successful, but their leader, Maj. Gen. W. F. Smith, cautiously paused, awaiting reinforcements. So it was Lee who got his reinforcements there first to halt the Union advance and set the stage for the longest siege in North American history. For historians of armed conflict, Petersburg with all its assembled military might packed into such a small area of disputed countryside was also the precursor to the grinding, sodden trench warfare that would characterize the First World War. With feints and sharpshooters picking opponents off, both sides sat there for nine months while the South hemmorrhaged. The North even brought up "The Dictator," a 17,000-pound mortar on a railcar. From a distance of nearly three miles, it fired two-hundred-pound shells into the South's fortifications.

One of the most dramatic incidents of the Civil War occurred in midsummer of 1864, when a regiment of Pennsylvania miners finished digging a five-hundred-foot-long tunnel beneath the Southern defenses at Petersburg. This may have been the last time the United States Army assigned experts to their area of expertise. It also involved an ingenious bit of ventilation with an underground fire and a homemade wooden pipe creating a draft that eliminated the need for telltale air holes. Northern troops packed the T-shaped underground cavern with four tons of black powder. About 110 artillery pieces and 54 mortars were primed and ready to help widen the breach. Three generals had drawn straws to see whose units would be first into the smoke, it having been decided that the original plan to send a Negro unit would not look too good if they suffered heavy casualties. The troops were massed. At 3:15 A.M. on July 30, 1864, the fifteen-minute fuse was lit. And at 3:30 A.M., nothing happened.

When the fuse had not set off anything by 4:15, Northern experts surmised that something was wrong. Henry Rees and Jacob Douty crawled back down the tunnel and, sure enough, the fuse had gone out. They relit it and got out themselves. At 4:45, in a thunderous explosion, the earth opened, spewing tons of dirt, pieces of 130 men, cannons, and guns high in the air.

So immense was the explosion that Union troops who had been assembled for the surprise assault were sent into flight—the wrong way. Regrouping consumed ten minutes. They charged. And when they got to the crater—170 feet long, 60 feet wide, and 30 feet deep—they stopped in awe and milled around. This blocked the breach for their own charging reinforcements, who, in turn, began oohing and aahing and milling about. And by the time the Union troops got organized, so were the Southerners, who retook their old lines. The crater is still there.

The skirmishes and attrition continued all autumn and winter, with Grant seeking to pound the city while surrounding it and stretching Lee's defenses, all of which the future president did. The North was helped, of course, by its massive undamaged industrial might but also by a thin piece of wire that the units trailed after themselves and stretched back to Washington, D.C. The wires allowed unprecedented cooperation and coordination between maneuvering units. Thus, like so many inventions that would later change civilian life, did the telegraph, which Mr. Morse first strung up the road between Washington and Baltimore, achieve widespread acceptance through its military application.

In March 1865, Lee determined to flee south down the highway to regroup in the Carolinas, a late-winter migration from the pressures of Northern life that many golfers now make. The Carolinas and Georgia had been undergoing their own military distractions, thanks to Grant's drinking buddy, General Sherman. But Lee's breakout from Petersburg failed. On April 1, General Philip Sheridan's men smashed George Pickett's Confederate forces at Five Forks and seized the Southside Railroad. Lee's right flank began to crumble on April 2. Southern stalling actions enabled Lee to evacuate Petersburg that night and avoid hand-to-hand street fighting in that already-battered city. But the end of the Confederacy seemed so near in those early days of April that President Lincoln was in the Petersburg area for two weeks for what would be his last meeting with Grant before the president's appointment at Ford's Theater. After a week's rearguard skirmishing during a hasty retreat to the west toward Lynchburg, the opposing sides met at Appomattox Courthouse to sign historic documents, which did not include the rights agreement for a future PBS documentary.

All that was left then was to clean up the battlegrounds. It was a gruesome, little-known task that did not really begin for one year, although plans had been made long before the carnage occurred. Anticipating casualties, though perhaps not the more than 618,000 of them that did occur on both sides, Congress had passed legislation in July of 1862, authorizing President Lincoln to purchase land for cemeteries "for the soldiers who shall die in the service of their country."

During the chaos of conflict, if men were buried, they were buried where they fell individually, in mass graves, or just outside the field hospitals where most of the wounded eventually died anyway from infection. In 1865, one commission surveyed the Petersburg area alone and found ninety-five separate burial sites for somewhere around five thousand Union troops. Most of the 30,000 Petersburg Confederate fatalities were buried in mass graves, by state, at the old Blandford Church on Crater Road.

The summer of 1866 was called "a time of searching" locally. Rewards were offered for every Union body, five dollars for any collections of bones with a skull. Some diaries recalled residents digging up anything or anyone that could qualify, not all of them heroes or soldiers. All the digging, opening up of remains, and resulting ill humors caused an often-fatal dysentery to sweep the countryside that summer.

The burial unit involved a line of one hundred soldiers walking at arm's length through the woods, across the fields, over the trenches, and along the roads and paths. The bodies had been buried in ditches, in mass graves, sometimes just in shallow depressions that left the toes and skull sticking out. Some had simple headboards; most did not. When a grave was found, often just by the appearance of disturbed earth, the entire line would stop until other workers arrived to collect the remains. The best-preserved, workers recalled, were the bodies from tightly packed trenches; those in coffins were mostly just dust and bones.

Many were incomplete torsos, a leg or arm here or there. Each body or portion was placed in a pine box and taken to the new Poplar Grove National Cemetery nearby. The burial unit had to range far afield in its conscientious collecting, and it accidentally brought in thirty-six incorrect bodies, it being impossible by then to tell what was loyal Union and what was faithless Rebel. This all took the better part of three years. But all the known remains were reburied.

Eventually, Poplar Grove contained 6,178 bodies, only 2,139 of them identified. One of them, grave No. 4841, was Billy Montgomery, an eighteen-year-old Pennsylvania boy who was shot at Appomattox just before the surrender on April 9. He didn't die for eighteen more days, but he is considered the last enlisted battle death in Virginia in the Civil War. At the Blandford Church cemetery, only 2,025 of the Rebels were identifiable. Around the crater, none of the 646 bodies could be identified. It was a very painful time in the region's history, although not so painful that the glorious defeat can't be reenacted with real gunpowder for the tourists on some summer Sundays.

In 1931, near the crater, someone found more bones, which led to some digging which led to twenty-nine more former humans being added to the array of tombstones at Poplar Grove just off Route 1.

But even though no one could know who was where, every spring soon after the United States were reunited, people began gathering at Poplar Grove for prayers and remembrance. There would be many other bodies found and lost and unidentified in subsequent American conflicts. But it was this simple local Petersburg observance that evolved into the nation's Memorial Day, which is now marked by car races, swimsuit sales, and special weekend golfing packages at the burgeoning resorts just down U.S. 1 in North Carolina.

Esse Quam Videri

DRIVERS ON U.S. 1 arrive in the Tarheel state right after ducking under Interstate 85 and just before passing a stern but puzzling official warning: STATE LAW BURN HEADLIGHTS WHEN USING WINDSHIELD WIPERS. Indeed, there are many signs along the highway, among them even signs for signs (BILLBOARDS SERVE AMERICA).

But there are also obvious and subtle signs of change. Billboards for teen-pregnancy services frequently show a troubled young female with the words *We Care,* and a phone number. Other signs announce Alcoholics Anonymous group meetings SUNDAYS AT 8. And in Henderson, North Carolina, is the impressive sign for the handsome First United Methodist Church, C. Clyde Tucker, Minister.

The congregation has 1,350 members now, a few less than last year and a few more than next year. "It's a pretty steady decline," says Reverend Tucker. The younger members are simply moving off in search of a future elsewhere. The older members are simply dying off.

A Virginia native, Reverend Tucker has seen it both ways during his thirty-eight years of ministering. Most recently, he had a modest-sized congregation in suburban Raleigh, where younger families predominate and the elderly parents have stayed back home in the countryside. He had over three hundred children in his suburban Sunday school classes, barely one hundred in rural Henderson. "Raleigh," he says, "is where the young couples go from here."

There are significant differences overseeing congregations in these places. Younger families, for instance, have less money, so a $250,000 fund drive like Henderson's to replace old wiring and plaster was doomed before it began. Younger parish members today have obviously shorter attention spans; the minister can see them start to move uncomfortably after ten, at most fifteen, minutes of sermoning. They begin to wiggle in their seats, adjust their ties, look around for something or someone to occupy their thoughts. Their favorite ploy is to clear their throats unnecessarily and sneak a glimpse at their watches in the process. Older churchgoers, some of whom portage their own cushions for the hard wooden pews, believe that fidgeting is impolite to the minister. So they don't check their watches; they just nod off at times. "I appreciate that," says the minister with a smile. "So I try to talk quietly so I don't wake them." That is accomplished by a blast of organ music that pops many heads straight up.

So even a veteran minister must learn to tailor both the length and content of his sermons to make them shorter and more obviously relevant to the problems and pressures of families with two working adults who are still in financial straits. In Henderson, Reverend Tucker can stress more religion and take maybe twenty or twenty-five minutes to finish the point he has been developing in his own mind since Wednesday.

Still, there are some advantages. Reverend Tucker gets fewer young adults at Sunday school classes than he used to, maybe twenty now on a good Sabbath. But they are an eager twenty, and an open twenty. "Oh, my, are they open!" he says. They need little encouragement to plunge right into deep, sometimes pained discussions of drugs and alcohol and sexuality and specific current problems with their mates. "I don't need to do much lecturing today," he says.

After the services, the congregants file quietly out the door and drive away on 1 past M & M Day Care, Clem's Auto Sales, Kay's Car Beauty Salon in an old gas station, and the Henderson Glass Co. (GIVE US YOUR NEXT BREAK). Traffic is usually light and almost always involves a few pickup trucks pulling a battered blunt-front aluminum boat for some sacred bass fishing outside of most every town, including Henderson. As chartered, Henderson the city should actually have encompassed nearly every foot of U.S. 1. That's because the North Carolina legislature approved the charter in 1841 before discovering a typographical error on the bill; it said the town (named for Leonard Henderson, a state chief justice) should cover all the land within a 1,200-mile radius; a clerk later corrected the official copy to 1,200 yards.

"Now my soul hath elbow-room," Shakespeare had King John say. And while John is not known to have traveled on U.S. 1, this area is the first truly rural countryside the highway has seen since before L. L. Bean's crowded parking lot in Freeport, Maine. This is not the kind of open Interstate scar drawn by a draughtsman's stainless-steel ruler beneath fluorescent tubes in a contractor's design room somewhere, running straighter than an arrow from here to there and beyond with anything taller than wild grass shorn from the face of the earth in a rigid war zone for traffic. Nor is the road here the kind of littered urban alley that sneaks into dim corners you'd really rather not know about.

No, U.S. 1 from southern Virginia down past Karen Barker's house and Jimmy Bennett's junkyard to the top of Florida and that Callahan Diner, where Chicago mobsters almost certainly probably maybe had breakfast, is a different kind of road. It is an old-fashioned country thoroughfare, largely free of man-made visual static, the kind of gently winding, often shady, and sometimes dipping roadway that invites the eye to look and the mind to wonder what waits just around the next bend, every bend. Or what sits back in that grove of trees peeking out, and wondering who sits in those passing cars? Or what on earth has that big old gnarled behemoth seen pass beneath its branches just three feet from the pavement's crumbling edge?

It is not a prosperous countryside, the reddening soil of the Southeast so heavily sprinkled with long-needle pines. There are abandoned tenant farms and barns overgrown with the lush kudzu vine that was introduced farther south decades ago to combat roadside erosion and now grows everywhere, even swallowing whole barns and following wires everywhere overhead. The Spectrum (HORNY LADIES—MASSAGE—ALL GIRL STAFF) near Franklinton doesn't use kudzu vines. But it has thoughtfully erected a tall fence to screen the view of its patron parking lot from the small-town eyes of any curious and gossipy passersby.

North Carolina's major roadsides also carry signs proclaiming that the next mile or two belong to some Rotary or Kiwanis Club, some women's group, or Scout pack, or even one family. They have adopted this section of state highway. They undergo safety training at the hands of Miss Betsy Powell. And they go every week or two and patrol their mile or two of roadway, combating the thoughtlessness of motorists who toss trash from their vehicles. The program, spreading in other states, too, was born in April of 1988. Since then, more than seven thousand groups have adopted more than 15,000 miles of North Carolina highway, with fifty-one groups caring for 102 miles of U.S. 1. The state provides plastic bags. The volunteers provide the bended backs. Road crews haul away the refuse to landfills, tons of it every year. And everyone feels a little cleaner for a few weeks.

Someone stole the antilitter program in New York City. But elsewhere, Americans have been fighting roadside refuse for decades, most recently through these highway-adoption programs, one small way that some seek to combat the widespread notion that government is responsible for everything that individuals don't want to do. In my childhood, our neighborhood sought to take care of itself with one neighbor, perhaps, mediating a dispute between two others. Police were called only as a last resort (you had to dial seven digits for help back then, not just 911). Likewise, each household was responsible for the roadside in front.

Later, even in the country, my father and I would patrol our property front several times a year, our gloved hands picking up a bag or two of sodden tissues, newspaper pages, tin cans, and faded boxes and bags. My mother said it looked much better afterward, but Dad told me it was really to collect the glass bottles, which, when unknowingly fed to our lawn mower, became like shrapnel.

Judging by the roadsides to this day, virtually every living soul in America litters, despite decades of public-service campaigns and thousands of signs threatening hundreds of dollars in fines. I know Florida uses convicts to collect roadside litter. But I've never actually met a convicted litterbug, or even an accused one. Litterbugs in America are like Nazi soldiers after V-E Day; every one of them served only on the Eastern front, which doesn't explain who was doing all that shooting back on the Western front. But it does show that when it comes to certain human behavior, everyone questioned happened to be somewhere else at the time.

U.S. 1 passes through the community of Wake Forest and by the college of the same name, which was founded in the 1830s in the forest of Wake. Early students at the Baptist institution were required to bring two sheets, two towels, an ax, and a hoe to help break ground. A century later, school fathers still frowned on any campus activity as wild as dancing. So the young men took the nation's main highway sixteen miles or so south to the big city (and state capital) of Raleigh.

Named for Sir Walter Raleigh, the city was designed in 1792. After decades of having a mobile state capital according to the vagaries of politicians and military enemies, the legislation specified that Raleigh be "an unalterable seat of government." (This may have contributed, too, to the state motto: "To be rather than to seem.") The legislation also required that the capital be within ten miles of Isaac Hunter's tavern.

Raleigh is home to numerous institutions of higher learning, including North Carolina State, where they have classes as well as a basketball team. With Durham (Duke) and Chapel Hill (University of North Carolina) within thirty miles, it has become a main focus of the famous research triangle. (It's really Bob Newhart's fault, but I can never hear the word *Raleigh* without thinking of that comedian's routine about Sir Walter excitedly telephoning his London backers with word of the discovery of snuff: Yes, yes, it's great. . . . What? . . . Oh, you stick it up your nose. . . . Yes, your nose. . . . Well, it makes you sneeze.)

With the modern-day snuff market weakened and with government deregulation and the development of regional airline hubs, Raleigh's economic-development officials, like those of Virginia, Florida, and Maine (and even those in places like Pittsburgh), have also sought to make the state and city into major tourist destinations, though Charlotte to the west got the professional bas-

ketball team, the Hornets. Although U.S. 1 flies by Raleigh as part of an encircling beltway, the attractive city offers much history and many old homes, which survive in part because the city simply surrendered to Sherman, who, unlike many Northern generals, was not a hesitant man.

North Carolina, which had 30,000 freed slaves when the Civil War began, was not eager for the conflict, but it did not shy from its share after Fort Sumter. The state had eighty-four Civil War battles and skirmishes on its territory and gave 125,000 recruits and 40,000 bodies to the cause. And if by war's end the North Carolina boys had had just about enough of those uppity Virginians giving orders everywhere, it was General Lee himself who produced the Tarheel nickname; see, North Carolina's troops saw themselves as admirably slow to retreat and claimed this was because they had so much of the state's native tar on their heels. So General Lee, who was headed to North Carolina when forced to surrender, said, "God bless the tarheel boys."

Raleigh also has an aggressive promotion campaign that involves bringing travel writers from as far away as Canada to tell their readers of the wonders of North Carolina (it produced three presidents—Andrew Johnson, James Polk, and Andrew Jackson, although Tennessee is trying to horn in on the Old Hickory attraction). North Carolina produced the writer Tom Wolfe (no, not *that* one, the first one). Authorities also feel the need to aggressively promote understanding of tourists among local citizens, handing out badges (ASK ME ABOUT RALEIGH!) and instant city reference guides to provide information and history. As one result, tourism and conventions bring more than $6.5 billion into the state each year, $1.5 million *per day* in Wake County alone. Already the state's third-largest industry, tourism will soon be number one.

From the airport, the tourists with golf bags tucked into the trunk of their rental cars head south on 1, which is called the Claude E. Pope Memorial Highway after a state secretary of commerce. With four lanes, it has limited access for a ways with exits such as Pea Ridge Road near the Haw River (are we in the South yet or what?). Pet owners take 1 enroute to the Wires Animal Hospital and along a chunk of 1 that the Triple A Roofing Company has adopted. Man-made ponds abound and very sandy soil. And then many trees, mailboxes, and front doors sprout that unusual modern symbol of loyalty, the wide yellow ribbon.

Such adornments do not indicate any great attachment to Tony Orlando. They usually mean two things: There's an American military operation going on somewhere in the world and a military base is not too far from that mailbox. In this case, the base is the sprawling Fort Bragg Military Reservation, whose western end touches Route 1 at Southern Pines.

Southern Pines, Whispering Pines, Pinehurst. Pinebluff. One might think there were a lot of pines around. There are. It's the state tree. (Turpentine, pines, and tar were big-time early industries.) And an exhortatory billboard confirms some problems: LET'S PUT THE HEAT ON WOODS ARSONISTS.

But in many places, the trees have made room for grass, lots of grass, more grass than most people can imagine planting let alone mowing, none of it from New Jersey's defunct sod farms. Across these miles of grass, thousands of people pay good money to walk on spiked shoes and ride in carts wearing striped awnings. The people are usually wearing one glove. They look down at the ground a good deal. They hit a little white ball. They walk after it. And they hit it again and again, until the little thing seeks shelter in a pond or a hole with a flag in it. Then the people do the same thing seventeen more times—and have a drink to celebrate.

Surrounding Moore County has thirty (!) golf courses. But in case this might prove inadequate, another eight are on the drawing boards. They are located in North Carolina's Sandhills region, 1,500 square miles of sandy soil and moderate climate created for golfers a few million years ago by a violent shift in the ocean floor combined with erosion and volcanic ash from the Appalachian Mountains. When the waters receded, they left vast quantities of sand at some of the highest altitudes of any American beaches. That combined with a moderate climate (average snowfall—four inches; average rainfall—forty-five inches; mean temperature—sixty-two) create a large area seemingly useless unless you were a migrating bird or a stagecoach operator who didn't like to stop on his way to somewhere else.

There weren't even any battles with the Indians; they'd long since given up on the so-called pine barrens, which had no livestock forage, and had gone into the western mountains. Eventually, the Cherokees were pushed even farther west into Oklahoma, though only two-thirds made it all the way. The first couple of white coastal settlements couldn't even get a scrap going between themselves; they simply disappeared in the 1580s. So, issues of mortal combat had to be imported; Scottish immigrants were loyal to the British king and local Revolutionaries weren't. So they killed and tortured each other. And that had to about do it for hostility venting, until that Illinois president started talking about military conscription to fight other Southerners over slaves.

Then in 1895, along came James W. Tufts, one of those rich Northerners (soda fountain equipment) who, long before Walt Disney, had just the idea to implant success in a stunted local area somewhere else. His idea was a health resort named Pinehurst for middle-income Bostonians tired of that city's long winters. The attraction for visitors was ease of transportation (a straight shot on the railroad) and, for him, the cheap real estate (one dollar an acre for cut-over timberlands). The founder of neighboring Southern Pines was John T. Patrick, who also brought in journalists to see the sights. He gave away free lots to northern doctors to encourage an undiluted recommendation of Southern Pines to their consumptive patients. It worked.

The idea of adding golf courses came soon after, reportedly when area farmers complained to Mr. Tufts that his guests were beaning their cows with little white balls. So Tufts hired Donald Ross, a sturdy Scot who apprenticed at St. Andrews and began designing American courses.

The area has had ups and downs and even abandonments to military uses. But the development in recent decades of a critical mass of affluent Americans with ample vacation time and disposable income fueled the resorts' resurgence. Today, more than a half-million people per year visit the county, creating jobs for over 4,500 others and leaving behind more than a quarter-billion dollars. None of North Carolina's ninety-two inland counties relies more on tourism. The visitors help support many amenities otherwise unaffordable to locals. To lure these visitors, local resorts have vacation packages for Memorial Day, July Fourth, and Christmas, Easter Family Getaways (egg hunts and croquet), Valentine Lovers Packages (candlelight dinners and carriage rides), Couples' Escape, tennis and golf schools, and Labor Day Wine-Golf Packages.

So popular has the area (and North Carolina) become that many visitors have decided to stay, or return to retire. While most American rural areas lost population in recent years, all the states along the Eastern Seaboard from Delaware through Florida, with their more moderate climates and generally safe and less-crowded cities, are among the fifteen fastest-growing states. Of Moore County's 18,500 households, one-third have just two members, and 7,100 of them have arrived in the last five years, which doesn't hurt the value of the land that nobody would even fight over.

At the moment, such in-migration of healthy, affluent senior citizens is welcomed by many areas; generally conservative and not eager for massive developments (once *their* condo units are complete), these people put few strains on public services such as schools. They provide a welcome cadre of business and civic experience and volunteer labor, not to mention a pool of conscientious part-time labor and off-hours customers who flock to the Early-Bird 4:00 to 6:00 P.M. specials at local restaurants.

Few experts can predict what the later local impact of this population sector will be as inflation continues to erode its fixed income from savings and as declining health creates demands for costly health and nursing-home care. But they can predict (and the nation's politics are already feeling) the power of America's gray panthers. Americans over age fifty make up 25.7 percent of the population now. By 2010, they will constitute 34 percent. And by 2020, after the baby boomers have tried to get some of the money back from all those Yuppie years of Social Security withholdings, about forty of every hundred Americans will be over fifty. Already today, the American Association of Retired Persons has over 33 million members. Like the AAA, 1 million more join every year.

One of those senior citizens is James Grissom, who is sixty-five and lives with his wife, Nora Lee, and 80,000 bees just a few feet off of Route 1 in Vass. Jim was a town policeman for thirty-some years. Then he retired to work as a part-time court bailiff over in Carthage and a full-time beekeeper out back. "He loves working with those bees," says Nora Lee. "Those little critters are right cute from a distance. But I don't mess with them."

Jim's been pretty busy as a bailiff the last couple of years, ever since all the thieving troubles started growing as the drug problem spread into the countryside. And the police, reflecting the taxpayers' fears and frustrations, began taking the drug people to court. Jim has to be careful, though, that he doesn't earn too much money to threaten his Social Security benefits. So the rest of his time goes into bees.

"Oh," said Jim, when he got back from work, "I don't have but twenty hives." But with 4,000 buzzing little critters in each hive, that's 80,000 altogether. Ten hives are out back; another ten are outside of town.

Beekeeping has nothing to do with the North Carolina professional basketball team called the Hornets (they got their name because, marching through here to end all that annoying American Revolution business once and for all, Lord Cornwallis was constantly bedeviled by guerrilla troops who harassed the British out of Charlotte and the future state but not before the good Lord wrote in his diary that this colony is "a veritable nest of hornets").

Beekeeping, on the other hand, is a $150-million-a-year industry in the United States, producing 100,000 tons of honey, about one-twentieth of a ton from Jim Grissom's hives. But, of course, the import of bees goes far beyond jars on a grocery shelf. In their endless search for blossoms, these winged workaholics pollinate American crops valued in excess of $20 billion. Caring for these busy little pollinators who have no business flying aerodynamically is an absolutely vital industry, but a threatened one.

Two kinds of tiny mites have invaded North America from Europe, thanks to jet planes that accidentally ferry them from continent to continent. And these mites are proving fatal to millions of American bees, who do not have an immunity, diplomatic or otherwise. One mite feeds off bee larvae, producing smaller, weakened bees more susceptible to disease. The other mite invades a bee's breathing tubes and sucks its body fluids. (Who looks in there and figures these things out, anyway?)

And then, of course, there are the "killer bees," the unusually aggressive species of bully bee, which was accidentally freed from a South American lab some years ago (just exactly why, pray tell, was someone trying to develop an extremely aggressive species of bee?). These insect instigators have been working their way up the Panama Highway ever since. (Why don't they ever go the other way? Where is it written that killer bees, fire ants, and bee mites should head for the southeastern United States, with all these golfers?)

The hope is that crossbreeding of the angry killer bees from South America with their more sedate cousins from North America, where revolution has been over a while, will produce a less aggressive blend. And scientists have now imported a group of mite-resistant queens specially bred by Brother Adam, a British monk who has raised bees for seventy-five of his ninety-two years. For unknown reasons, Brother Adam's bees are rarely infected with mites. The hope is that, although they come from a strict religious upbringing, the British queens will fool around with North American drones and produce succeeding generations of stronger offspring.

But all that is more complicated than Jim Grissom's work. "I've been with bees since way back yonder," he says. "I helped my mother with bees back in the thirties." Jim wears all the right gear and he smokes up the hives to calm them down when it's "honey-robbing time." He still gets stung once in a while. "You mash one and he gets annoyed. Especially on damp days. I would, too."

He gets maybe ninety-six pounds of honey in a good year. The bees produce more, but Jim leaves a good supply to help them through the sometimes surprisingly chilly North Carolina winters. At times, he feeds them extra sugar water. Twice a year, Bill Shepard, the state bee inspector, drops by to check on all the little Grissoms (the two Grissom boys are both grown and, of course, moved away, one to become a minister supervising an orphanage and the other working in state government).

Jim admires his bees' dedication and industriousness. "You go out there in the winter," he says, "and they're all huddled around the queen, keeping her warm." The drones, of course, give their lives to mate with the queen, who couldn't care less if it is good for them. The worker bees wear out their little wings in a short life of hovering and portaging pollen back and forth across the highway for a few days.

Every day, they range out up to a three-mile radius from their home hive, hauling back load after load in a zigzagging patternless flight path that inevitably ends at the correct hive. Even with ten identical-looking hives in the same yard—and then more working a farmer's fields out of town—the workers get confused about only one thing. If someone moves a hive more than a few feet from its standard location after the bees leave for work, the bees will come home to the old spot, hover, and fly around a small area searching in vain until they die of exhaustion.

So Jim looks out for the bees. And he sells some of their annual production (five dollars a quart with the comb, four without) to friends and neighbors. And every morning as they sit down to breakfast and thank God for their continuing good health, Jim and Nora Lee Grissom quietly appreciate the bees' good work as they dollop one spoonful of honey each onto their bowls of Special K.

The insects do not seem bothered by all the vehicles on busy route 1, which include local cars and pickup trucks from the many tobacco farms, delivery trucks to the new grocery store, a few through trucks from the carpet plants, golfers, of course, and, twice a year, thousands of race fans.

People have been racing each other since well before the first Olympics. Americans learn to race as soon as they walk: Race ya to the car! Last one in is it! Competition seizes even strangers at stoplights or passing joggers. Americans race on foot, on bikes, on snowmobiles, in planes, in boats, in sacks, and in wheelbarrows. They race eating pies and, of course, in cars, all kinds of cars. When I was little, at least one summer Saturday every year, my father and I would go to the car races in a gritty little industrial city at an old dirt track with walls made of wooden slats painted to advertise motor oil and soft drinks.

There was popcorn and soda pop to consume on those warm carefree evenings, which was attraction enough for me. Sure, the slivers from the worn wooden benches could easily penetrate your pants. But what were you doing wearing good clothes to the car races, anyway? After that first Saturday, I was hooked: the noise, the smells, the dust, the excitement of competition, the vicarious victories. These were not famous folks out on that track in those old jallopies. These were other dads and sons. They came around the fourth turn in pairs, side by side, twenty or more in a pack, the motors accelerating. My eyes darted from the front row to the starter to the front row to the starter, who jumped up high and waved a green flag as if he'd just become rich.

Then, as if one worn leather boot jammed down on a single bare metal pedal for twenty cars, the rumble grew.

They leapt ahead as one, well, two; my favorite blue car had stalled. But in an instant the pack had passed. "Oh, wow!" I yelled. Dad could not hear me amid the roar of the engines and the crowd and the fleeing fumes.

Dads—my dad, anyway—didn't yell. But he was excited; I could tell. He sat up straight. He rubbed his hands and cleared his throat and smiled down at me. And then they came back around again. The roar was wonderful. The power. They blasted by. Whooee! The air was filled with smoke and dirt. It was great. Here they come again, spreading out some now, as the pack did in life, too.

I liked them to pass me in continuous packs. It kept the roar going longer, so strong that it shook my hairless little chest. The power in those machines was amazing. They were decrepit and old, I know now, and the fathers had fixed them just the night before with a fair dose of cusswords and perhaps a part stolen from the family car in the light of a bare bulb on the cool cement floor of their oily garages. But this was big-time competition. Here they came again. It was so wonderfully loud. And when each driver backed off the accelerator just before the first turn, the motors protested with a loud series of rumbling backfires. *BLAM! BLAM! Bleh, bleh, bleh, blam.* Each explosion sent little blue flames flickering out the side. Then the feet were on the gas again. The roar built. Around and around. Time and again. Until the checkered flag.

We could catch our breath then and maybe a nearly warm hot dog. And then they were off to the races again. All evening. Beneath the lights. It was so exciting. I had no idea who won. We knew no names. Couldn't hear them over the underpowered P.A. system, anyway. Who cared? It was the dramatic competition. Afterward, we could walk across the track right where they had just been racing. I got chills just standing there. There in the dirt infield stood the weekend warriors, knocking back a beer, with their other arm around a worshipful woman, talking to friends, reliving the excitement, the could-have-beens, the if-onlys. The resting cars were still warm to walk by, the soil oily beneath my PF Flyers. Once, I found a broken flange of metal on the ground. It was still warm, too. Amazing that so many had walked right by such a treasure without seeing it. But Dad said I could keep it. I stashed it in my pocket, then kept it on my dresser for a long while.

"What's this?" my mother would say, as if something dirty and oily had sneaked into her home.

"I don't know," I said.

"Well, then, let's get rid of it, shall we?"

"No! We can't. I got it at the track with Dad."

"What are you going to do with it?"

"Nothing. Just keep it."

"Well, if you're not going to do anything with it, let's get rid of it."

"No, Mom. I might use it when I make my own car."

"Well, don't put it in a drawer or somewhere clean."

Mothers. They don't know anything. Who would put the nucleus of their first racing car into a drawer of underwear?

They race differently, of course, in different cultures. The Europeans sit on the grass savoring fine cheese and sipping some wine with a swell bouquet and watch millionaires in fire-retardant suits wheel their hybrid land rockets around an intricate road course that places a premium on changing gears and finesse passing at sleek speeds that may reach 120 miles an hour. Americans, however, have refined this automobile-racing business into a high-speed, beer-guzzling, color television–gripping science of automotive engineering, merchandising, sex, advertising, and stomach-jumping noise where wine has no place and cheese is something that the good old boys cut when they hit the men's room under the main grandstand.

Stock-car racing has nothing to do with stock cars, of course. Any resemblance between these fine-tuned honeys packed into sticker-coated fiberglass and the family Lumina with the dog hairs in the backseat is purely mental. But it does feed the "me, too" mentality of many Americans who will pay a fortune to identify themselves with a winner, even an orange car carrying the five-foot

letters of a washing detergent around in circles for two or three hours of competition. And twice a year—once in March and again in October—the attention of American stock car racing fans is focused on a small city in southern North Carolina, Rockingham.

It is a 210-year-old community named for another Frenchman who liked the Colonies. Once, it was the focus of corn, cotton, tobacco, and melon farming. Still, it's the Richmond County seat. A few miles north of town is a paved oval, 1.017 miles around, where grown men with the most wonderful collection of names in sports make about two thousand left turns at speeds upwards of 160 miles an hour in desperate pursuit of a few hundred thousand dollars, some points in the NASCAR driver's competition, and fleeting fame on the Monday-morning sports pages.

The National Association for Stock Car Auto Racing is a far cry from the ramshackle, dirty T-shirt show of my youth. It is a spectacle made for American television—cars, color, speed, danger, big money, commercialism, adoring fans, and small-town helmeted heroes named Neil Bonnett, Lake Speed, Sterling Marlin, Dick Trickle, Ricky Rudd, Rusty Wallace, and Morgan Shepherd. Previous years have goldened the competitive memories of the likes of Fred Lorenzen, the Allison brothers, Bobby Isaac, Cale Yarborough, Richard Petty, and, of course, Curtis Turner, who despite broken ribs won his last Grand National event at Rockingham in a classic clash with Cale before a crowd that, if an informal survey is any indication, included just about every living soul in North Carolina that day.

The North Carolina Motor Speedway is a 189-acre site designed by computers and completed in 1965. It has grandstand seats for 38,000 fannies and infield space for 17,000 more hot-dog eaters. Across Route 1 is the Rockingham International Dragway, which doubles on race weekends as a landing strip for all the private planes that drop in to bask in the powerful automotive glamour and all their well-dressed passengers who've come to sell and be sold. For 321 days a year, the track stands idle and empty. On forty-two days, it is rented out for automotive testing, go-cart racing, and private car racing. But officials estimate that just the two five-hundred-mile NASCAR race weekends bring about $46 million into the area, more than the cotton and peach crops ever did in many years gone by.

Rockingham's race days are two of the twenty-nine events on the annual NASCAR Winston Cup Circuit that begins every February farther down 1 in Daytona Beach and ends every October in Atlanta, with intermediate stops at tracks from Arizona and California to Michigan and New York and back down through Dover, Delaware, and Richmond, Virginia.

Even if some snooty North Carolinians blame the racing crowds on Virginians who leave home for the weekend, North Carolina has three of the big-time tracks, more than any other state. This is widely credited by legend to the high-speed driving skills perfected by previous generations of good old boys who dodged the cars of revenuers in the state's mountains after North Carolina instituted prohibition in 1907. Canada's rumrunners sought to defeat authorities in Maine with guile and stealth. North Carolina's boys just outran the badges. Apex, back up 1 toward Raleigh, was the headquarters for the Baldwin gang, who brought the stuff in from Virginia. But there may have been one or two other people involved.

However, stock-car racing has gone legit big time now. NASCAR's twenty-nine Winston Cup races draw more than 3.1 million spectators a year. Competitions in the junior divisions of NASCAR bring in over 16 million more. The average fan, according to NASCAR figures, is male (62 percent), married (58 percent), holds an academic degree (91 percent graduated from high school, 41 percent went on to college), is in his thirties (31 percent), owns a home (64 percent) and at least two cars (69 percent), and earns more than $35,000 a year (48 percent).

And where that many working Americans gather with money and wheels, there is bound to be, you guessed it, megamerchandising. There are many millions of dollars spent on programs, hats, jackets, pennants, stickers, photos, foods, and liquid refreshments at the scene. But even those huge sums pale in comparsion to the vast sums car sponsors pay to associate themselves and their products with these 3,500-pound thoroughbred machines, creating the Skoal Bandit Oldsmobile, the Lifebuoy soap Chevrolet, the Zerex Ford, the Tide Chevrolet, the Miller High Life Buick, and the Bull's-Eye barbecue sauce Olds, among many others. These sponsorships change every year as drivers, sponsors, car builders, and mechanics jockey for the best winning position off the track as well as on. And each position change is covered in regional newspapers as if it had something to do with sports.

They Don't Shoot Horses Here, Do They?

AS IT DOES IN northern Maine, U.S. 1 in South Carolina passes through areas of countryside impoverished in terms of money but rich in tradition and family values. *Dum spiro spero,* says the state motto: While I breathe, I hope. The Spanish were the first whites to arrive, in the 1520s, on the coast, of course. In 1670, the British came to Charles Town. And the two nationalities began fighting. Although New England gets all the historic press, 140 engagements of the American Revolution were fought in South Carolina, which has long prized its independence.

When their concerns over eroding states' rights grew

too great, South Carolinians got up in arms, literally, and opened fire on Fort Sumter. South Carolina became the first state to secede, a position of prominence it paid dearly for four years later when General Sherman passed through. South Carolina didn't get back into the Union until 1868 and suffered through many decades of poverty long after. As late as 1927, when there were already forty motor vehicles for every mile of state highway in South Carolina, only 303 of the state's 4,951 miles of state roads were paved. Although highway construction has long since caught up with the other states, South Carolina's residents either remain frustrated or eat the fattiest foods possible because they have the highest death rate from heart disease of any state.

South Carolina's long, bleak economic period was brightened in recent times by the spread of diversified light industries (and four nuclear power plants) seeking cheap labor displaced by the decline of labor-intensive agriculture, and by the growth of tourism. Visitors have become the second-largest industry in a state with such a scenic seacoast from Myrtle Beach, which actively pursues wintertime Canadian tourists like Maine come summer, down through fashionable Hilton Head Island.

In Camden during the Revolution, British authorities hanged many local rebels. In Cheraw, up the main highway closer to the North Carolina line, General Nathanael Greene appointed a colonel named Charles Lynch as judge advocate to handle the legal entanglements of running a revolution. Judge Lynch performed his duties with such efficiency and dispatch, often foregoing time-consuming details, that his name became synonymous with trial-free executions.

Cheraw, first settled by a lethal Celtic stew of Welsh, Scots, and Irish, enjoyed some moments of prosperity as a shipping point when steampower brought maritime commerce up the Pee Dee River. It was a glamorous time, those falls, winters, and springs of steampower when the *Maid of Orleans,* the *Osceola,* the *Chesterfield,* and the *Swan* pulled in after another four-day round-trip to the sea (the boats were laid up all summer due to low water). When

the faster railroad came in, of course, it wiped out the steamboat, so the waterfront became a landfill.

Cheraw also had a novel idea of punishment, which did not cure public drunkenness but did provide a legal legacy of shade; the city required every convicted drunk to go into adjacent woods and bring back a tree to plant along city streets. Many of them still survive (the trees).

There have been a lot of problems over bridges in Cheraw what with accidental fires and intentional war. Advance elements of Sherman's boys encountered there the rear guard of Hardee's troops retreating toward Fayetteville. That was also the only engagement of the Confederate gunboat, the *Pee Dee,* which fired some shots and was scuttled, never having been anywhere but Cheraw. Hardee burned the bridge, which slowed the Union troops and got a southern chain of fast-food restaurants named for that Confederate leader, though they are owned by Canadians.

None of this Pilgrim stuff for McBee, South Carolina. It was founded at a barbecue in 1900, where lots were auctioned off by the Seaboard Air Line Railroad, whose demand for wooden ties created a local lumber industry. Although barely 800 people live in town today, another 200,000 reside in the environs, a fair number of them retired couples.

But not all the seasonal travelers following the route of U.S. 1 are human. Ronald Snider presides over the Carolina Sandhills National Wildlife Refuge, which has made McBee a major rest area for the East Coast's winged winter commuters. Some 46,000 acres of grass and pine-covered hills of sand, once North America's seacoast, present an ideal refuge for some four hundred species of waterfowl and migratory birds, including the endangered red-cockaded woodpecker. Not that so many red-blooded hunters are blasting red-cockaded woodpeckers out of the sky, although they do ping away at many moving things; it's a more insidious social destruction of the kind of landscape the creatures need to survive and procreate. For the four-inch-tall woodpeckers, it's old-growth pine forests.

McBee has perhaps 120 colonies of these beautiful birds, each colony consisting of a nesting female, her mate, and up to six other males. The extra males can provide ready suitors, of course, should the main mate disappear. But their main purpose in the meantime is as extra caretakers of the young, helping to round up sufficient sustenance for the babies as nature does its best to build the population back up. The colonies of woodpeckers each live in their own self-assigned territory, moving around in a half-dozen tree cavities they have helped hollow out.

The refuge on U.S. 1 is also a popular rest area for migrating songbirds. The little nuthatches will stream in by the thousands on a Monday, and by Wednesday they're all gone. They are joined in their noisy October and April visits by warblers, junkos, robins, quail, and, of course, geese. These large birds come sailing in from nothern Canada by the hundreds, all a-honk. They wander the ground awkwardly. But up above, they are gorgeously graceful, moving powerfully through the air at speeds of forty miles an hour or more. Each bird takes its restful turn at the outer tip of the squadron's V, assisted slightly there by the uplifting air currents passing off the ends of its peers' wings just ahead and to the side.

Springtime is the busiest time for birds at the McBee refuge, with thousands of winged creatures chattering and flying about, refreshed by a warm winter's rest, satiated on tropical Florida's insect population, and emboldened by nature's carefully timed urge to multiply. The migrating birds are followed by hundreds of migrating bird-watchers in their seasonal uniforms, complete with binoculars dangling around their necks. The wheeled coastal commuters seem to prefer the local highway over Interstate 95, off to the east on a more direct route to the sun. The bird-watchers arrive in flocks barely behind the animals. Just like the birds they are watching, the bird-watchers favor certain parts of the refuge.

The deer hunters favor other areas. They can hunt with bows and arrows for six days, with old-fashioned black powder guns for five days, and with modern weapons for just seventy-two hours. "I used to hunt deer," says Ron. "But as I've gotten older, I've gotten to where I'd much rather watch 'em, if you know what I mean." The refuge manager lives down 1 barely five minutes from work and there's hardly any traffic when he's on the road early, which is no reflection on his driving.

The population of refuge visitors, though still small by the standards of other major natural attractions such as Freeport's factory outlets, has grown in the last decade, as if, Ron wonders, people are savoring more their dwindling wilderness. Nonetheless, he and others make regular refuge checks at all hours, even after closing, because there are poachers around who like the idea of a stuffed woodpecker sitting on a mantel. The refuge closes parts of its hiking trails in winter, too, though the three dozen ponds remain popular with bass fisher-persons.

Ron has worked for the Fish and Wildlife Service in Mississippi, Arkansas, Florida, and next door in North Carolina and Georgia. With a boy who chose an indoor career (computer engineering) at Clemson and a daughter in junior high, he'll be working for some years yet, but has decided to join the area's retirement population at some point. "You know," he says, "I just hope my bosses don't find me down here for a while."

In some Northern cities, outdoors is where you must go on the way to somewhere important, unless there's an air-conditioned walkway to save that bother. In South Carolina, the outdoors is important. Gardens are prolific in the climate, not surprising given its rare frosts and eight-month growing season. Plants include colorful magnolias, azaleas, rhododendrons, and, of course, gardenias, which were invented by and named for Alexander Garden, a physician in Charleston. With typical South Carolina modesty, coastal Charleston describes itself as the place where the Ashley and Cooper rivers merge to form the Atlantic Ocean.

The Spanish were in South Carolina first, establishing the settlement of Santa Elena in 1566, where the Marine Corps's Parris Island training base is now. The French got involved in the state's coastal hostilities at times, as did the pirate Blackbeard and the English, of course, who did not like the often sunny climate. The area has, however, proved eminently hospitable for horses. First introduced by the Spanish, the horse wreaked terror on Central American natives. Equine escapees formed the nucleus for the surviving wild-horse population of the American West. And hundreds of their pampered thoroughbred descendants now live in Camden, South Carolina, on U.S. 1.

First called Fredericksburg and then Pine Tree Hill and, finally, Camden, the community was named for Charles Pratt, Lord Camden, a champion of Colonial rights in Britain's Parliament. Camden, the state's oldest inland city, is another one of the towns linked by U.S. 1 along the fall line, the height of navigation on area rivers. It was settled in 1732, though its real growth began in the 1750s with the arrival of enterprising Irish Quakers and Joseph Kershaw, a merchant who gave his name to the county after being deported by the British to Bermuda, where people pay a lot of money to go nowadays. Andrew Jackson is said to have been jailed in the British headquarters in Camden for refusing to polish the boots of a British officer. And, according to local lore, the only reason Sherman spared Camden's surviving Civil War homes was because it was raining so hard that day that burning homes was difficult.

But there is nothing imprisoning, nearly seven hundred miles out of New York City, about the area's immense lush green fields gently surrounded by white rail fences. There, not unlike professional baseball players several hundred miles down the road in Florida, thoroughbreds frolic in the fields during breaks in their winter training. Three racetracks, a steeplechase run, and more than two hundred miles of bridle paths honeycomb the Camden area where horse shows, horse races, fox hunts, and polo are popular and provide significant tourism income. (Boykin spaniel breeding and training are also popular, but they're for hunting ducks and doves, not riding.) None of this, when combined with the proximity of Interstate 20, hurts the area's sales efforts to attract more

jobs for the county's 45,000 residents. The last census revealed an exodus from rural America; more than half the rural counties lost population, a total of 2 million residents gone from a million square miles of geography. Camden's Kershaw County, however, is one of the rural counties that gained population; the common characteristics of the gainers: ocean views, mountain resorts, or, as in Kershaw, farmland within an hour's commute of an urban area. This has made the county's median age (thirty-two) unusually low.

With a mean annual temperature in the sixties, the area's winter equine population is nearing two thousand; hundreds of others winter farther down 1 in Aiken, South Carolina. But even those good-sized communities of horses are a far cry from their ancestors' herds. With horses providing the pulling power for farms and the transportation in both countryside and cities, the horse population of the United States in 1902 was 17 million—this at a time when the number of motor vehicles was 23,000.

Indeed, in the early days, motorized cars and trucks were hailed as the cure for a major pollutant that threatened the health of American society. The dangerous pollution was horse manure. The old books don't say who was doing the measuring, but they do estimate that the country's population of horses dropped 2 million pounds of manure and 60,000 gallons of urine on American streets *every day*. Not to mention the problem of disposing of the weighty carcasses of horses who died at work.

No question the motor vehicle took care of the urban horse-manure problem. For a few locations where a couple dozen horse-drawn carriages remain as tourist attractions, the ever-vigilant city councils, which are too busy to address the modern problem of automotive gridlock, did handle the lingering horse-manure problem swiftly by requiring the animals to wear a sack beneath their rear. Thank goodness for such bold leadership in facing up to the horse lobby. That leaves only the problems of crime, social decay, and canine manure to be confronted.

Of course, in the countryside, manure is a resource

recycled onto the land, like cities do unsuccessfully with their plastic bottles. It is a revealing symbol of the changing times that the United States Department of Agriculture (with Washington offices just off U.S. 1) keeps close track of our population of hogs (55.5 million) and cattle (99.5 million). But it no longer knows how many horses there are. Counting horses is a job left to the Department of Census, which has enough trouble counting people who don't want to be counted, but still can compute that 2,456,951 horses call America home. That's up nearly a quarter-million in five years. And since most of them are pets or decoration, it is also a measure of the country's affluence, and perhaps some lingering nostalgia.

Karen Barker is another resident of Camden who has spent most of her life living on U.S. 1 but has yet to enjoy much affluence. She was fourteen when she met her future ex-husband. She was fifteen when they married. "I threatened to run away unless Mama signed for me," she recalls. They got married on August 10. Three weeks later, Karen was old enough to drive.

"I thought I was in love," she says. "You know, like teenagers do." A year later, of course, there was a baby, Jamie, who is now older than his mother was when she got married. To his mother's relief, he has stayed in school. Later came Shon and now Dustin, the baby. Karen was married for fourteen years altogether—much of that time in Florida, Alabama, and Virginia, where her husband worked construction. Now, after a long separation, a divorce, and a lengthy court fight over child-support payments, she says she remains friends with her ex-husband.

She got the trailer in the divorce and sold it for the down payment on her tidy three-bedroom home just north of downtown Camden on Route 1. Karen has four sisters and five brothers, several of them carpenters. So they helped build it. "I just had to have my own house," she explains. And in return, Karen helps her brothers, especially Tony, who runs a prospering catfish farm.

The days are long. Karen is up at 5:00 A.M. Shon is off to school at 5:30, Jamie at 7:00. They must walk across some neighbors' lots to another street because the school bus will not stop on the main highway. Soon after, she drops the baby off with a relative and drives her fifteen-year-old Malibu to work down at the Hot Spot convenience store or the Wal-Mart or, more often, to clean fish out at Tony's place. Some days, Karen and the other girls at Tony's slice up eight or nine thousand fish. A few go to area restaurants, where Karen does not order them because she knows what they look like raw. But most are boxed up. A truck comes every couple of days. And the country fish go off to feed some faraway city folks; Karen isn't sure where, but the truck turns south on U.S. 1.

The boys are home by late afternoon. By seven, Karen has cooked up supper. Shon's in bed by nine. Jamie stays up as late as he wants. "He's sixteen," says Karen, "You can't tell them what to do no more." So she often leaves him on the couch and cuts off the TV when she goes to bed at eleven.

There are some good times, too, though. Every year there's a family reunion, mostly from her mama's mama's side. So many relatives come to town that they must gather in a state park. And Karen has met Bruce, well, more than met him. Bruce is the baby's father. He helps with money, too, and the two have discussed marriage someday. He's got three kids from a former marriage. They live with him. They don't see their mother much, so Karen ends up mothering them, too, sometimes.

Bruce lives with his mother, but Karen talks to him daily. There's an occasional Friday night visit to the dance hall just up the highway. And on some nice weekends, they take Bruce's camper out by the river for a night.

Like everyone, Karen's got her regrets. "I wished I'd a stayed in school," she says. "I would have never gotten married that young. But that's just things, you know, I went through."

With forty-seven inches of precipitation per year (and all of it rain), the yard floods easily. So Karen must leave her car up on the road. But she was lucky when the hurricane went through. Many trees fell down, but they all missed her house. And Karen thinks one of these days, given the new possibilities for women in modern society, she might go get a construction job. "I could be a helper for eight or nine dollars an hour," she says. Although that would mean leaving home by 5:00 A.M.

Karen also sometimes finds herself thinking about her father, who died when she was six. "I don't remember anything about him," Karen says firmly. "But every once in a while I'll be doing something, cuddling the baby or driving somewhere, and I'll get a flash. I'll see Daddy. He always looks the same. He's smiling." Karen's daddy was an auto mechanic. He died in a car crash—on U.S. 1.

The highway leaves Camden gently, passing Roy Kelley's Used Car Lot Auto Sales in a former gas station, the high school, where the wild winds bent over the stadium lights, Smith Road, where Kids Kare day care is routinely registering new toddler students, and out into the countryside, where it is called the Old National Highway. With such light traffic and the rural roadside so near, U.S. 1 in South Carolina offers the pace and place for real bird songs to seep through the open car window, the thrilling trills startling the urban driver more accustomed to horns and grinding gears.

Near Columbia, the road passes the Pontiac Day Care center and Children's World and then Robert Alexander's Spring Valley Animal Hospital. There is, too, the South's ubiquitous Waffle House, one of the many chains that are not so slowly standardizing all the menus and restaurant architectures.

These waffle emporiums are almost always near an Interstate. They have the same tile floors, the same booths, counter stools, and, out front, the same newspaper-dispensing boxes. These twenty-four-hour breakfast places have the same groups of round-the-clock regulars, who appear at precisely the same time every day or night, exchange the same jokes and jibes about the weather and local sports teams, and, I swear, have the same trio of friendly waitresses in every single Waffle House. I don't know how that trio of women gets around so much and knows in advance which Waffle House I'm going to patronize next. But they are a modern-day com-

mercial wonder, and, I suspect, received their training at the Duncan Hines Academy. I could do without the diners' aluminum-foil ashtrays. However, the waffles are delicious and, unlike most chains, are cooked the way I request. Imagine that, food cooked to order in this day and age. (Sometimes, when you have a half hour to kill and your blood pressure has been running a little low, try ordering a cheese quarter-pounder without ketchup at McDonald's; the number of incorrect variations are infinite.) But most times in mid-meal at Waffle House, I must stop to determine precisely where I am.

Sherman knew where he was, precisely. Columbia was laid out as the state's compromise capital within three miles of South Carolina's exact geographic center. And while, once again, U.S. 1 moves vaguely through town, the streets were built in a precise checkerboard pattern and 150 feet wide because this was believed to inhibit the spread of malaria, which does not often appear in the Chamber of Commerce brochures.

The state capital with its huge dome and some of the largest monolithic granite columns in the world was begun in 1855. Fortunately, it was not complete by February 17, 1865, when Bill Sherman's boys torched 1,386 buildings on eighty-four blocks downtown. Everything went up except the French consulate (more interest paid on Lafayette's goodwill) and the capital, which only had walls at the time. Metal stars on the west and southwest walls mark where Union shells hit, the kinds of scars some Southerners savor.

Columbia was also the site for drafting the state's Ordinance of Secession on December 17, 1860, at the First Baptist Church. Little Woody Wilson, the first president to take the United States into a world war, did much of his growing up over on Hampton Street. And George Washington, who never did travel abroad and spoke nothing but English, came through on his southern tour in 1791, noting everything in his diary in presidential prose noncolorized by ghostwriters. George left, however, as have numerous other Columbians. About two-thirds of all state capitals gained population in recent years; those that got smaller were mainly in the East—Richmond; Harrisburg, Trenton; Albany; Charleston, West Virginia; Augusta, Maine; and Columbia, South Carolina.

Sunday, May 22, 1791—Rode about 21 miles to breakfast, and passing through the village of Granby just below the first falls in the Conagree (which was passed in a flat-bottomed boat at a Rope ferry,) I lodged at Columbia, the newly adopted Seat of the Government of South Carolina, about 3 miles from it, on the North side of the River, and 27 from my breakfasting stage.

The whole Road from Augusta to Columbia is a pine barren of the worst sort, being hilly as well as poor. This circumstance added to the distance, length of the stages, want of water and heat of the day, foundered one of my horses very badly.

Beyond Granby 4 miles I was met by several gentlemen of that place and Wynnsborough; and on the banks of the River on the North side by a number of others, who escorted me to Columbia.

Monday, May 23, 1791—Dined at a public dinner in the State house with a number of Gentlemen and Ladies of the Town of Columbia, and Country round about to the amount of more than 150, of which 50 or 60 were of the latter.

Tuesday, May 24, 1791—The condition of my foundered horse obliged me to remain in this place, contrary to my intention, this day also. Columbia is laid out upon a large scale; but, in my opinion, had better been placed on the River below the falls. It is now an uncleared wood, with very few houses in it, and those all wooden ones.

Even then, most American communities looked the same—small and wooden. In Washington's time, the nation had barely 4 million people and only six cities with more than 6,000 residents. Thirty-five years later, travel had increased so much that stagecoaches ran from Co-

lumbia to four points at least once a week—Greenville, Charleston, Camden, and Augusta, Georgia. Typically, any plans to build turnpikes and canals died with the advent of the railroad, and the arrival of the automobile nearly a century later proved the downfall of Columbia's streetcars.

U.S. 1 in front of the state capitol is, of course, called Jefferson Davis Highway. Heading out of town to the southwest, it becomes the Augusta Highway. By any name, it is long since familiar: a collection of separate stores and aging strip malls, interrupted by fast-food outlets, used tire stores, billboards (ABORTION STOPS A BEATING HEART), Millie the Psychic Reader and Advisor, video-game parlors and batting cages, some abandoned houses and abandoned cars, and the occasional church (NO NATION IS STRONGER THAN ITS HOMES). In towns such as Lexington and Batesburg (a major source of the country's caskets), the old road is so much a part of local life and so close to its roadside residents that travelers can smell Sunday suppers simmering. In the countryside, the houses are back from the road, but the mailboxes are close by, some of them decorated with painted flowers or ducks or dogs or, in the case of Andy Storey, proudly proclaiming his allegiance to Clemson with a bright orange mailbox carrying the school's trademark tiger pawprint.

Aiken was a railroad town vigorously defended against the evil you-know-who by the South's General Joe Wheeler. It has long been a winter training ground for horses, and late last century, it began developing as a resort for winter-weary Northerners. U.S. 1 there, however, does not pass through the loveliest part of town. It ducks by the Kozy Kort Motel (OPEN 24 HOURS), a couple of auto salvage yards with rusting hulks piled high, Tacky Tony's Fireworks, and another prison, the twenty-two-acre Aiken Youth Correctional Center built in 1975 to handle 252 troubled youngsters. Today, the jail houses 320. Then with no formal announcement, the highway crosses into Georgia past the Alamo Plaza Apts. and Motel (AMERICAN-OWNED) and briefly becomes a six-lane expressway with metal barriers.

Jefferson Davis Had Breakfast Here

IF THERE IS ONE thing that can make a few modern-day Southerners angrier than remembering what General Sherman did to their region, it's what he did not do to some towns like Augusta. He did not burn it or sack it or attack it. He didn't even shoot at it. He went right on by, as if he didn't think the place was very important. A fair number of Augustans reacted like spurned suitors, furious that that Yankee didn't even think their hometown was worth plundering.

Historical drawings never do generals justice. Whether it's Alexander or Marc Antony, these gentlemen always appear in freshly pressed togas, looking as if the hairdresser just left their field tent. With the development of photography during the Civil War and before the advent of airbrushing, we get a compellingly detailed look and feel for what soldiers and generals really looked like. Most look as if they must have smelled. But looking at General Sherman's photo, his hawkish nose, his piercing eyes, his small head, his bow tie wadded up and turned askew, his several days' growth of beard, one immediately wonders, How did so many Civil War generals manage to shave enough to avoid a bushy beard but still look so bristly around the chin? Was there a secret blend of Burma-Shave or a now-defunct make of shaver that trimmed ten days of whiskers down to, say, three. This would be designed to keep facial hair under control but still preserve a stylish unkemptness implying whatever dirty chins and wispy cheeks are supposed to imply? Some modern-day professional players of hockey and football seem to keep such devices in their lockers.

Anyway, reading the chronicles of a stern Sherman—make that Mr. Sherman—one is impressed with his forthrightness: "My aim then was, to whip the rebels, to humble their pride, to follow them to their inmost recesses, and make them fear and dread us." And looking at his fierce face, especially the strangely menacing shock of short hair, I, for one, would not seek the assignment of awakening the general from his afternoon nap. Nor would I be likely to complain to him about his not pillaging my place.

But being overlooked by devastation has long annoyed Augusta, which was named for a German princess who married the Prince of Wales. Augusta (the city) had a major arsenal that should have turned the city into a prime target and created a forest of historical markers. When his former West Point roommate, an Augusta native, wrote Sherman fifteen years after the war and asked why he hadn't smoked Augusta, the general replied something to the effect: If you're that disappointed, maybe I can come back and burn it. By then, however, the profitable sprouts of tourism could be seen as rich Northerners began to discover Augusta as a winter haven. Eventually, Augusta attracted the parents of a man who would become president (Woodrow Wilson), a man who acted like he was president (John D. Rockefeller), and a vacationing general who commanded more men in combat than any man in history, and then became president (Dwight Eisenhower). During the second week of every April, Augusta's 400,000-plus population is supplemented by a few dozen professional golfers swinging clubs in and ten thousand spectators spending $35 million at the Masters Golf Tournament.

But getting back to the most important issue, one favorite local theory for Sherman's snub noted that as a young soldier, Sherman had been assigned to Augusta, and that he had a close West Point friend from there who had a sister with a lovely southern name, which no doubt made her pretty, which may have led to a romance, which might have produced a baby, who probably would have died, given conditions in those days. Such an infant would have been buried in a small cemetery not far from the arsenal. And if all that was true, theorized people who did not know the general personally, then Sherman, overcome by an infectious southern sense of gentility, probably would not want to attack his own son's cemetery.

The fact that Sherman never did marry is seen by some as further proof of the poor man's irreparably broken heart. Others see this as further proof of irreparably romantic thinking, which comes along with large flowers and a climate that produces peanuts and pecans.

The simplest explanation, which rarely sells newspapers and novels, may be best. Sherman didn't attack Augusta because he didn't need to attack Augusta. He was in a hurry to blast his way up the seaboard and get this war over. Having gone through Atlanta like the wind and prepared the ground for that movie, he could leave further study and refinement of his tactics to later German military planners, who called them *blitzkrieg* and *schrecklichkeit*. Sherman could simply slip around Augusta to sack Savannah, shut the railroads and the Savannah River there, and seal a well-defended Augusta off to wither without supplies: no cotton out, no money in.

Georgia was a crucial state. Although it initially prohibited slavery, it later became a bulwark of that institution and actually accepted the last shipment of slaves to reach North America. Georgia was the site of the first major gold strike in North America (1828), the first use of ether as an anesthetic (1842, Dr. Crawford W. Long), and the first use of that crucial invention by that New Haven native, Eli Whitney. That invention, of course, has become known as the cotton gin, which is actually Georgian for "cotton engine."

Georgia might also have been known for inventing the sewing machine. The Reverend Frank R. Goulding spent years of his free time at home next to the Presbyterian Church just off Route 1 in Bath working on a machine that would use up lots of cotton very quickly. But the minister never thought to move the eye of the needle down near the sharp point. So Howe got the first patent, instead.

Georgia, however, remains the largest state east of the Mississippi. Once, it reached all the way from the Atlantic to the Mississippi. But for a penny and a half an acre, land speculators in league with unscrupulous but suddenly very wealthy Georgia legislators deeded away Alabama, Mississippi, and Tennessee. Later legislators, who had not benefited, and the U.S. Supreme Court declared that deal unconstitutional. However, everyone outside Georgia ignored the decision, anyway, which didn't get the land back but didn't hurt later Southeastern Conference football rivalries, either.

Georgia was not founded by Cherokees, who did not name it for George II. It was resettled by a group of private British investors headed by James Oglethorpe. The barrier islands offered super sites for pirates and for British forts to keep an eye on the Spanish, who were down in Florida looking for the Fountain of Youth and getting their overloaded treasure ships sunk in perfectly timed hurricanes that had to get along without a name for a couple of hundred more years. Georgia's substantial rivers offered cheap and efficient inland transportation. This, in turn, fertilized the flourishing plantation system and society for the folks who brought you institutions like the slave market and Andersonville Prison, which is not on U.S. 1.

Atlanta, that cosmopolitan New South city and Olympics site, is not on 1, either. In fact, not much is on the 220-odd miles of U.S. 1 in Georgia. South of Augusta, the Georgia communities of size are confined to one: Waycross. The others remain sunny outposts of a fading Southern agricultural era that invisibly produces peanuts, pecans, sweet Vidalia onions, and surplus crops of women with middle names and men with two first names. This same large landscape and sense of family and small-town values also helps Georgia produce the sort of homespun loneliness and down-home romances that have inspired so many country-music songs and their singers. Although by the time most of the singers have gotten the money from their fans for the tapes of the nostalgic music that means so much to both, they are all stranded in suburbs many miles away from their rural roots. The ones left behind can hear it on the radio, though. And the sense of loss is still genuine.

South out of Augusta, U.S. 1 follows the path of an old Indian trail and almost immediately crosses another famous byway, which Erskine Caldwell traveled with his father, the Presbyterian minister, before turning Tobacco Road into *Tobacco Road*. The thoroughfare is called Tobacco Road because that's where the mules hauled the hogsheads of drying leaves to the loading docks on the Savannah just below Augusta.

The four lanes of 1 are divided by a grassy median and road reconstruction equipment as they move along, before narrowing to two lanes (three up the hills) near Boggy Gut Creek. The trees are few; the homes, too. The fields are flat and large, their surviving fences cluttered with windblown balls of cotton that look to be taking refuge there. In Wrens, Caldwell's boyhood hometown, the federal highway stops its southwesterly wandering and turns southeast for the last time. It passes a throng gathering in black at the M. W. Calloway Funeral Home just beyond earshot of the excited shrieks of continuing life in a nearby touch-football game.

Wrens acquired its nonapostrophed name from John Wren, who acquired the land in a trade for two blind horses. Over the years of swings in the agricultural economy, there are those who have thought that perhaps the new owner of two blind horses got the best part of that deal. On the east side of town was Pope Hill, the stagecoach stop where Union Army guards allowed the new former president of the Confederacy, Jefferson Davis, to have breakfast.

Down the road stands Louisville, like Augusta and Milledgeville, former state capitals. As late as the 1930s, Louisville's old slave market still stood, topped by a large bell. The bell was a pre-Revolutionary gift from the king of France to the city of New Orleans. The bell was borrowed by pirates and sold in Savannah to Louisville merchants, who needed a good noisemaker to summon slave buyers and warn of Indian attacks. The old Louisville City Cemetery contains the grave of Governor Herschel V. Johnson, who joined the nation's forgotten ranks of the politically obscure by being a losing vice-presidential candidate. (But even Hannibal Hamlin, the victor in that vice-presidential contest, was destined to remain obscure: After only one term, he was replaced on the 1864 Republican ticket of the doomed Abraham Lincoln, thus missing the presidency by only a few weeks.)

Unlike parts of Virginia, where pieces of 1 are so little used that short grasses grow in the cracks of the pavement, U.S. 1 through Georgia is smooth—today. Once, it was not so. Legal scholars exploring the lost early days of the U.S. Supreme Court, when records were not kept or have been destroyed, have uncovered letters from the often-discouraged and -exhausted federal justices referring to this area's difficult travel conditions. Although most people believe John Marshall was the nation's first chief justice, he did not take office until 1801 and was, in fact, the fourth chief justice. Before his thirty-four-year reign, in the court's early days, the justices did not sit in the nation's capital and receive legal supplicants. They rode a geographic circuit on horseback. In 1798, the lack of roads and an intimidating flush Georgia swamp with chest-deep water kept Justice James Iredell from reaching Savannah. In 1793, after just six months on the Supreme Court, Justice Thomas Johnson sent President Washington his resignation. "I cannot resolve," the justice wrote, "to spend six Months in the Year of the few I may have left from my Family, on Roads at Taverns chiefly and often in Situations where the most moderate Desires are disappointed." Despite many requests to Congress, the justices continued to make the circuits until the modern federal court system was established in 1891.

Today, even nonlawyers would find a fine highway, and it's still the main road anywhere. The open spaces, the emptiness of the farmland, even the airiness of the woods encourage speeds that may exceed the legal limit. The countryside, however, is green and grassy and, were it not for all the Baptist churches, might be mistaken for the Midwest. Each town has a modest motel, in case someone decides to stay some night. The stores are few and far between; whatever they sell, many also offer ammo so everyone can protect their own constitutional rights and be ready to protect the peace through force. The stores also typically sell a limited array of music cassette tapes wrapped in dusty cellophane that seem to carry an awful lot more tape than music these days, like the big aspirin bottles that turn out to be half-filled with cotton. But at least they're not like those weirdo rock groups with the braids who get up at concerts and just pretend to sing; didja hear about that?

Just about every town has a used-car lot, too, and, increasingly, at least one old-folks home, the young folks no longer able to provide local care because they are no longer living locally.

It was these rural areas of the United States, laced throughout every region, that for so long seeded American cities and spawned their suburbs with the ideas, values, ambitions, and proprieties of a small-town social system where people behaved as if what others thought of them meant something. These models were never as pocket-size perfect as Jeffersonian ideals had envisioned, and some people who were very imaginative or just downright strange found the system perhaps boring or impossibly confining or maybe both.

So two or three generations of Americans liberated themselves by moving away into rented urban spaces with different rules of mass survival and walls that allowed them to hear neighbors arguing and flushing and doing other things. And they felt freer for a while on the concrete, no longer being harnessed to the demanding land, and less confined. Although as time went on, they needed considerably more door locks, and alarms on their car, and maybe a small can of Mace on their key chain like a bad-luck charm.

A powerful sense of drift, nostalgia, and wistfulness seeped into the music then, as the strength of old values waned, as the supply of hometown humans shriveled, and as the manners, mores, and morals of that distant small-town society began to seem dusty and quaint, at best, and possibly even old-fashioned—which is not helpful in a commercial society that so prizes youthfulness and "newer than new" over anything reeking of old. It was probably just a coincidence, then, that "country" became fashionable and city country-music stations boasted they had many more listeners than country country stations. They saw commercial opportunity in that, not a contradiction.

For those who have not moved away yet, Swainsboro has a stock-car track where the same guys and dates and the same families in T-shirts tend to gather on warm Friday nights to mark the end of another long work week and the beginning of another short paycheck. Some driveways are marked with half-buried tractor tires painted white. Others have ramshackle tables of faded wood offering homegrown pecans. Most every front porch has at least one chair sitting, waiting for a sitter. Many lawns seem to have produced car skeletons; they sit there silently, waiting for rust to claim their soul or an inspired mechanic, whichever comes first. Dogs are everywhere. So is the scent of hay and, after a good rain, the promising perfume of wet soil. Although, of course, many fields go unplanted now.

Incongruously, just north of Baxley on the southern banks of the Altamaha River sits an immense structure. It is the Edwin Hatch Nuclear Generating Plant, one of two nuclear facilities owned by the Southern Company, a southeastern electrical conglomerate. The 810 megawatts and the 820 megawatts produced in Hatch's two reactors comprise 11 percent of the company's power generation (hydroelectricity is 3 and coal-fired generators are 86). Hatch's power is poured into the state's electrical pool, which tends to flow north now, where so many of the people have gone with their hair dryers, microwaves, TVs, and rechargeable screwdrivers that save so much twisting of the hands.

But to the south a ways, nearly eight hundred miles out of Washington, D.C., where hurried pedestrians and harassed motorists exchange urban finger signals, sit Alma and Dixie Union, in a town where some pedestrians take the time to watch a passing car on Route 1. Then comes the aptly named community of Waycross, Georgia.

In 1818, settlers began gathering around Waycross, building blockhouses for protection from the Indians. This kind of adversity often encouraged organized religion and many believe the city's name came from the expression *Way of the Cross.* As late as the mid 1930s, one book noted that Waycross was home to "15 white and 24 Negro churches." Others believe the name emerged from the location being the eventual convergence of nine railroads and five highways. As late as 1870, barely fifty people lived in Waycross. Some Waycross natives have left, people like Pernell ("Trapper John") Roberts, Ossie Davis, and some guy named Burt Reynolds, who played football down in Florida. Today, thanks in part to the continuing development of that adjacent Florida market, the city has about 20,000 residents, an aged air of prosperity, and hopes of mining the vein of tourism. And it wouldn't mind being a little more famous for a more famous little citizen who never really lived there.

One of those lifelong residents is Jim Bennett, Jimmy B, as some friends call him. Jimmy has nothing to do with the area's big bee industry or pecans or them tourists who drive so slow because they think U.S. 1 is not an Interstate and they don't know the highway troopers the way the good old local boys do. Jimmy has nothing to do with the railway, either, although his daddy did; Will Bennett died early, leaving the other boy and the two girls grown up and gone but four-year-old Jimmy alone at home with his mom. She worked in a local grocery. Jim helped informally, so his mom got a little more salary. She rented out their own house to cover the mortgage and the mother and son shared a room in their own house. "Hey!" Jimmy says. "When you come up poor, you learn all kindsa stuff in case you'll need it."

One of those things he learned was to talk right fast for a southern boy. He quit school at sixteen, did a hitch in the air force, and went to auctioneer's school up Kansas City-way. As his speaking speed and general incoherence increased, Jim found hisself in demand locally. He was announcer out at the racetrack, rattling off lineups and lead changes and winners' names over the rumble of misbehaving engines and nonexistent mufflers. He loved cars, anyway. They gave lots of country boys a real sense of power and satisfaction just driving and fixing 'em up. Maybe they couldn't control the loss of jobs, the railroads' decline, and all the other people who did graduate moving away to the cities and getting newer cars. But, by golly, they sure could control how this old car ran and looked. Nothing baffling about motors if you looked hard enough or listened close enough. They could tell where the bad bearing was or when the timing chain was about to go. And they'd fix it themselves. Maybe chrome the manifolds. And then they'd wear their banged knuckles and oily shirts as badges of courage and conviction in a Wal-Mart society where everyone still started the same day the same way, pulling their jeans on one leg at a time.

Jimmy B did his announcing on weekends. Weekdays, he was a blur up and down U.S. 1 auctioneering cars down in Daytona on Tuesday, in Orlando on Wednesday, Jacksonville on Thursday, and over in Valdosta, Georgia, on Fridays. "I tell you, we covered some miles in those days," he says as if the sixties and seventies were ancient history. But he can still talk right quick. And he proves it, launching his fluid tongue into the sale of a fictitious Buick: "HeyIgottwennysebenhunnerttwennysebenhunnertherenowtwennyeighttwennyeighttwillyagot twennyninetwennyninehunnerttwennyninehunertYes! twennynineIgottwennyninehowboutthreegimmethree nowhowbouttwennyninefifty?Twennyninefifty?Igot twennyninethentwennyninelastchancetwennyninehunnertoncetwennyhunnertwice SOLD to you there with the handsome mustache for twennynine hunnert dollars."

Then he sticks his frayed cigar back in his mouth with the wad of Levi Garrett and almost smiles.

Jimmy still keeps his mouth in auctioneering part-time. Like many raised as only children, he enjoys the attention focused on him being up front. And he doesn't mind the money. The city hired him recently to get rid of some old stuff. With his 9 percent commission, Jimmy picked up nearly twennyninehunnert dollars for a few hours of talking the audience into parting with $32,000 for old machinery and fireman's uniforms. The city turned its junk into money. And a lotta folks had some seemingly free entertainment and then went home with a bargain, even if they had not arrived at the sale driven by the need to acquire a fireman's old rubber coat.

After auctioneering cars, Jimmy got hisself a used-car lot in Waycross, which made him a little money—"Sellin' cars is no chairahtee, boy, y'heah?" Selling kept him around cars—"I'm a car freak ever since I can remember, working on someone's car sometimes for free just to learn how they worked." And the lot let him race his own stock cars, what he calls round-trackers. Everyone in town knew that four, maybe five nights a week Jimmy B and the boys could be found in, on, or under some old car, adjusting, installing, or fixing something from the last encounter with someone else's left-rear fender.

He got into restoring old cars then, too. Old cars are in now, even outside the country, so you can talk some serious money with those babies. Take Jim's '36 Ford Cabriolet convertible. It was a piece of crap when he bought the rusting hulk out in Salt Lake City. Now, it's a piece of beauty, worth maybe thirty grand, but Jim'll never know cause he'll never sell that baby, although you say the word *forty* and he might start thinking, although that baby means an awful lot.

"This ole car had no door handles. So I had to research what they looked like and then go find some. Didja ever try to find a coupla door handles fifty years later? Now I think I know so much, but I didn't know they had two models that year, a regular 1936 model and a late 1936 model. Instead of metal, the late model has a wood-grain dashboard. So I had to find one of those for it to be right. Then one day, I got ahead of myself. I thought

I was so smart, buying a chrome emergency brake handle for a hundred dollars. But then I come to discover the handles were just painted that year, so here I am sandblasting all the chrome off'n my new one-hundred-dollar chrome emergency-brake handle."

The new old cars look real good, but they make even better memories. "They always look old and awful at first," Jimmy says, caressing a door, but maybe you better let him be the only one to touch her without permission. "But you know in your mind what she can look like and you pour yourself into it and you sweat and you scrape your knuckles and get all dirty and you rebuild that sucker to look just like you want it, not like somebody else wants, but just like you like it and you take it to a show and you get first place. Now that just feels sooo good, y'heah?"

Prizing such pleasures, Jim feels a little out of step, though not so much that he plans to change. "Today, folks don't seem too willin' to work for stuff like that. They want to buy it and have it right now. Modern folks don't like to wait. Always in a hurry. Well, you can buy it good as new. That's for sure. But it'll never be the same as when you built something yourself. You know every bolt in it and you know its history and you made some of it. Like my Cabriolet. You know what this is? It's the back-seat outside. Know what they called it? The mother-in-law seat, 'cause you'd put her in here and at fifty miles an hour, you couldn't hear her inside."

Matter of fact, it was restoring cars that got Jim into his latest business trouble. It's not really any trouble. That's just the way guys talk when they love something, the word *love* not coming easily to such lips. Jim runs one of Route 1's auto salvage yards. One day, oh, twenty-three, maybe twenty-four years ago, Jimmy needed some part for an old Mustang he was fixing up. So naturally he went out just south of town to Huey Arnold's salvage yard—Huey, the boy, not the father. Huey looks up at one point and says, "Jimmy, instead of just one part, why don't I sell you the whole damn yard?"

Jimmy had kinda liked the place for some time—all those cars sitting, waiting to be restored to shiny life or sold, piece by piece. So he said, "What you got in mind for a price?" Huey said he'd check with big Huey, but Jimmy had best remember that when the old man gave a price, it was his bottom price to start. Jimmy said he remembered.

Next day, little Huey says, "How 'bout forty for everything?"

Jimmy, the auctioneer, moved quick. "Sold," he said.

The rest is history, though unwritten until now. Jimmy gave up round-tracking to run A & B Auto Parts. Jim doesn't ever advertise—word-of-mouth is still much better in small towns—but if he did, the A& would get him listed first in the Yellow Pages, which is not as first as AAMCO, but then this isn't Atlanta, either. The salvage yard was twenty acres of junk—WE SELL THE BEST— SCRAP THE REST. But some years ago the government came through and took some land to widen 1, which is sometimes called Memorial Drive in south Waycross. So Jim had to correct all his business cards: "18 Acres of Late Model Wrecks." Actually, as everyone knows, Jimmy's got some good old stuff, too. Keeps the cars—2,000, maybe 2,200 of them—arranged by model and year, oldest toward the back, where the grass is taller.

"Morning, Jim. How you?"

Jim nods. "Mornin', Judge."

"Y'all got a knob like this here for a '74 Matador?"

"I think I might. It's in the back row past the trees and the old Mack truck."

Geez, with 2,200 dead vehicles and the good Lord only knows how many knobs, wires, wheels, and whatnot sitting out there with flat tires looking like other people's thankfully forgotten junk, how does a fifty-three-year-old former round-tracker and part-time auctioneer remember everything without an IBM-PS2 with high-density expanded memory and overloaded interfacing?

"Well, lemme tell you, boy, if it determines whether you eat or not this week, it helps to focus your memory a whole lot, ya hear what I'm sayin'?"

Jim rules over his eighteen-acre domain with a cautious geniality that emerges unless you get a little city-pushy or try to telephone. Jim hates the telephone. Oh, he knows business needs one, dammit. And he often answers it eventually unless the caller gives up or someone else gets to the noisy phone first, which his friends, knowing Jimmy B, often do out of consideration for his hospitality. Jim tore an entire front seat out of one of his wrecks. And he placed the upholstered bench on the floor in front of his battered counter so the regulars can sit and swap stories and make up the audience for Jim, who reaches for the phone with the cigar still unlit but aging.

"A&B. . . . That's me. . . . Yup. . . . Speaking. . . . Yup. . . . It's possible. . . . I might. . . . Just south of town on the federal highway. . . . Around five." If the caller had any more questions, he's talking to the dial tone by now—which is better than the busy signal A&B gives off most times because Jim's had it with the damned ringing and has taken the phone off the hook and let it lie. "There," Jim says proudly, "that takes care of that racket."

People seem in such a hurry nowadays, not as personal as they used to be when folks thought that to be polite when they encountered each other, they had to chat a while, even before getting down to business. Especially before getting down to business. Now, if anyone younger asks after anything or anyone, like as not they're faking. You can see it in their eyes. Hear it in their voices. And them thinking they're so smart and smooth and in charge of these dumb ole boys who didn't finish high school. So if Jim knows you and your business, then you can go in his yard looking for the part on your own. If he doesn't, you don't. And if you're one of them that tries to save time by doing everything by phone, the hell with ya.

J.B., Jim's helper, knows his boss, though. In a few minutes, he'll come along and see the phone off the hook. "Hello? Hello?" J.B. says on the off-chance that Jim has left someone hanging for help. Then J.B. hangs up the phone. And almost instantly, it rings.

Jim didn't used to hate phones. He thought they were useful most times, until the modern-day generation took to calling him at home about business. About 1:00 A.M.

it'd ring, ring, ring, and some voice all friendly-like would say howdy, he needed a starter for a '67 Rambler. Would Jim just leap out of bed and run out to his salvage yard on the edge of town and find a twenty-year-old starter in the dark so the caller could get to work by 7:00 A.M.?

So Jim leaped from bed and tore his home phone out.

At first, it was a little inconvenient. But after a few days, it was wonderful, all peaceful-like at home. If he wants to make any calls, Jim keeps track and makes them from a pay phone at the Shoney's restaurant when he gathers with the other morning-coffee regulars. Those people he cares about know his routine; they can find Jim out at A&B by 10:00 or 10:30 six days a week, and they know better than to go looking by the house. "When I die," Jim says, "there's three things I hope they don't have where I go—phones, worn-out cars, and women."

He presides over his cluttered shop with the young women on the old calendars, the peanut-vending machine whose contents probably deserve carbon dating, and the Pepsi machine full of Coca-Cola. Jim knows his customers and he knows his sometimes-selfish Pepsi machine. So, out of consideration mainly for himself because he's tired of answering complaints when the falling coins do not cause the proper dispensing, Jim years ago scrawled a sign with an arrow and taped it in just the right spot. HIT HERE, it says.

The judge is back. He found the knob for the '74 Matador, right where Jimmy said. Damned right, Jimmy thinks. He says nothing, chews the cigar. The man explains he's trying to sell an old car and it's missing the knob off the radio, which doesn't really matter because the radio works fine. But every would-be customer points out the damned missing knob, as if they expected the man to drop his price a couple thousand. Jim knows that kind from way back. So now the Matador owner is agonna settle this knob problem once and for all.

"Whaddeye owe ya?" he asks, reaching for his wallet.

Slowly, Jim looks down at the little one-dollar knob, then back up at the former judge. Jim smiles.

"Have a real nice day, Hugh."

"I sure do thank ya, Jimmy B," says Hugh, silently agreeing to the contract. Word-of-mouth.

A kid comes in. Jim knows his daddy. The man has sent him to get six hubcap rims. Jim directs him to the rim assortment. The kid returns.

"How much, Jim?"

"Oh," says Jim. "Fivehunnert'll cover it." He waits for the shock to appear before smiling. "How about twenny?" That seems so much less than fivehunnert, even if it was more than his daddy told the kid to pay. Jim deposits the wrinkled twenty in the cash register, his wallet.

Jim says he doesn't work so hard anymore, but he's out at the yard pretty near every day, Monday through Friday and half of Saturday. He closes that afternoon to prepare his '79 Plymouth Arrow for the night's drag races up in Douglas.

When he watches those yellow lights blinking, his heart starts to pound. Life is so rich then, like a good cigar. The colors become vivid. Who needs drugs? The lights blink yellow. And the heart pounds. And they blink yellow. And it pounds. And blink. Pound. Green! And Jim slams that pedal down and pops the clutch and it's as if he's in a rocket of his own making. Going straight and true. No sound, just the engine's roar. The crowd cheering, no doubt. No fucking phones. Flying down that racetrack, loving every hundredth of a second, in a car named "Just Havin' Fun." In 660 feet, he does 107 miles an hour. Takes his breath away. Damned fine. He'd say he loved it, if he could. Someday, he might make 108.

Sundays, Jim goes to the yard, too. Maybe to restore some of his old road signs or another antique for a museum he might open someday. Or to adjust his Plymouth for next Saturday. Maybe thumb through the weathered scrapbook of old Waycross postcards he keeps beneath the counter—old-fashioned colorized shots of the Plant Café, Plant Avenue before so many windows got boarded up, the old motorcycle cop what's-his-name, the old boys out at Hardy's Super Service, standing all proud and straight in front of the pumps they used to fill up all the travelers' cars before the Interstate went in along the coast. And then there's Jim's damned bookkeeping to do, too, you know. He hates numbers, unless they're on money or the side of cars.

Oh, and feed the guard dogs. The sign over the front counter says, THIS PLACE PROTECTED BY A PITBULL WITH AIDS. But they're really like small Dobermans. Actually, there's just one now. The male up and died a week ago. Right after Jim gave away the latest litter, wouldn't you know. The bitch—Jim calls her Girl—has been mooning around the yard ever since.

Weekdays, Jimmy B's at work. He lights up his cigar. A regular yells from the front. There's a customer waiting. "If I take long enough," Jim mutters, "maybe he'll go away." But he walks back toward the man he's never seen before in his life.

"Hey, Jim." says the man. "How y'all doin'?"

"It's Monday."

"Yeh, sure. You got any fourteen-inch tires worth a cuss?"

Jim spits out a little tobacco juice. The man's obviously in a hurry. Jim's not.

"Didja hear about the old-timer driving down the highway?" Jim asks. The stranger looks bewildered. He looks around at the audience of regulars. They're looking at Jim, who accepts the silent encouragement and continues.

"The old-timer sees a sign at the roadside park. It says, 'Free Picnic Tables.' When he gets back to town, his friend says, 'Didja have a nice picnic?'

" 'Yeah,' says the old-timer, 'but I near to broke my back getting that picnic table in my trunk.' " Everyone laughs. Hard. Save one. He's there to do business for Chrissakes. This Bennett guy is telling dumb jokes. But it's Bennett's place and the man needs the tires bad. So he nods, smiling faintly, like he's got a little gas this early in the morning. And everybody's still laughing. And Jimmy B is looking around at each member of the audience, smiling himself, and mentally raising the price

that this jerk is going to pay him by fifteen dollars, no, make it twenty.

The man looks like he's feeling better now. He's trying to get the conversation back on track before another joke hatches.

"Fourteen-inch wheels?" the man repeats.

"Well," says Jim all pleasant-like. "Let's go see." But Jim's not exactly running.

The two men walk off, the customer lost, watching the ground for droppings that might soil his good shoes. Looking desperate to fill the silence of their stroll, he eagerly explains why he needs fourteen-inch wheels— which Jim hadn't asked. Girl is tagging along for company. The phone is ringing.

Jimmy B looks back over his shoulder. He blows a puff of smoke. And he winks.

Lost on the edge of Waycross is the old U.S. One Drive-In. Someday in a century or two, its huge screen and its sprawling grounds with metal stanchions lining rows of dirt mounds may offer archaeologists revealing artifacts on what a society was like before it moved its entertainment indoors onto machines without names, just initials like TV and VCR. The drive-in is abandoned now except for one corner, which the county uses as a temporary dump and transfer station. It's not too difficult to envision the archaeologists of the future surveying this scene, rummaging through the remaining refuse, and then producing learned articles for prestigious New England journals on their theories of life in the 1950s and 1960s. These academic tracts would describe how members of this obviously primitive Georgia society apparently gathered quite often in a large open area. There, they worshiped a tall, flat screenlike structure pointing toward the sky and offered up strange sacrificial gifts of soup cans, cow bones, plastic motor-oil bottles, and waxed milk cartons.

Another primitive society still struggles to survive several miles south of Waycross. It is the legendary Okefenokee Swamp, six hundred square miles of lush overgrown wetlands and watery channels that isn't really a stagnant swamp at all. Fed by an average sixty inches of rain a year and several underground springs, the Okefenokee is its own constantly circulating complex water system that drains itself to form the headwaters for the Suwannee and St. Marys rivers.

An exotic ecosystem of tropical trees strung with moss, lush brush, watery grasslands, drifting lilies, and contorted cypress, the water actually purifies itself, in part using tannen soaked from cypress roots. The tannen turns the water a dark brown, which makes safe hiding for many creatures. An inch or more of leaves and vegetation falls to rot on the ground and in the waterways each year. Every three or four decades, in a timeless cycle of renewal that twentieth-century man sometimes tries to stall, several years of scarce moisture lead to low water and perhaps a lightning storm that ignites the accumulated natural refuse. Historically, the fires could burn for months, sometimes following subterranean veins of methane gas from rotting vegetation to burn beneath water channels and emerge to spread on the other side. Without the regular cleansing fires, the swamp would choke on itself, slowly silt up, become a marsh, and then a vast meadow.

Today, the Okefenokee is part of the National Wilderness System, with a nonprofit interpretation center run in one corner on state land. There, local residents like Lamar Hall show visitors some of the natural infrastructure such as the soapweed, which, when rubbed between wet hands, produces a natural soap. As he poles the small, flat-bottomed boat through the humid heat around twists and turns, bubbles of gas rise from the sodden peat beds below and a pair of bulbous eyes break the dark water's surface nearby, a baby alligator checking out the passersby. Lamar and the others try to convince visitors to look but leave the wilds alone, to drive through and enjoy it but not to veer their cars out of the way in order to run over snakes, which suffer from as poor a public perception as wolves once did. But we've taken care of the wolves.

Bears live in the swamp, and otters, deer, egrets, great blue herons, ospreys and owls, even water moccasins. Also resident on the center's grounds is Oscar, an alligator who is somewhere between twelve and fourteen feet long and 650 to 800 pounds; no one wants to get close enough to ask. Oscar hangs around the center's gate, sleeping mostly. Every week or two, the keepers toss him some bad pork or an unsold chicken or two from a local butcher. It keeps him less aggressive and less likely to chase visitors, which would do little for tourism. Oscar eats only seventy pounds of food a year. And he's perhaps only eighty years old, which is young for the Okefenokee.

Some of the cypress trees along the overgrown channels were seedlings when Christopher Columbus sought the bottom of U.S. 1. Columbus did not reach the isolated Okefenokee; his boat was too big. But other characters have, some of them fictitious and some of them stranger than fiction. Among the real ones, was Will Cox, a veteran guide who helped survey for the narrow-gauge railway that carried the lumberjacks into the swamp and the large cypress logs out. There was Obediah Barber, who lived in the swamp for ninety-two years and once fought off a bear with a chunk of firewood.

And then there was Lydia Smith Stone Crews, a tough-talking, hardworking, metal-mouthed woman who stood somewhere over six feet tall and weighed probably over two hundred pounds; no one wanted to get close enough to ask. A shrewd businesswoman who made considerable money by buying used timberland and letting it reseed, Miss Lydia outlived her first husband. So in 1928, at the age of sixty-three, she got herself a new man one-third her age. She called him Doll Baby and thought so much of J. Melton Crews that when he found himself in prison with a thirty-year sentence for murder, she bribed the warden with a large check to obtain her hubby's release. Then she stopped payment on the check. And since it would have been awkward for the warden to explain what he was doing accepting any check from a prisoner's wife, let alone a bad one, he could do nothing. Miss Lydia died in 1938, leaving a rich widower, but presumably taking along her purse-sized pistol, just in case.

The Okefenokee's most famous resident never really lived on U.S. 1. His name is Pogo. He was a cartoon character created by Walt Kelly, the Bridgeport native who did live along the highway back up in Connecticut. Kelly, once an animator for Walt Disney and later an editorial cartoonist for the New York *Star,* was casting about in the library of Dell Comics back in 1948, looking for a suitable swamp setting for a new character. More a fan of wry satire than violence, Kelly just happened to pick the Okefenokee for the backwater home of his snub-nosed swamp-philosopher possum and his band of well-informed animal intellects.

They were the forerunners for a new breed of political comic strip in the United States, long accustomed to cartoon panels with simplistic Dick Tracy stories or Dagwood pratfalls. Kelly's crew—fuzzy little Pogo, Albert Alligator, Churchy Lafemme, Beauregard Bugleboy, Porkypine, and other animals who bore a striking resemblance to political personages of the day—watched the day's news and commented satirically on it themselves. "Don't take life so serious," Pogo was fond of saying. "It ain't nohow permanent."

At the peak of "Pogo," in the late fifties and early sixties, the strip was read daily by millions in more than 450 newspapers. Although Kelly died in 1973, the impact of his cartooning lived on in a refined line of contemporary political cartoons such as Doonesbury. Pogo's oft-quoted line "We have met the enemy and he is us" is actually a corruption of the famous line after the Battle of Lake Erie attributed to another Route 1 native, Rhode Island's Oliver Hazard Perry.

In 1988, with the permission of the cartoonist's widow, Selby, "Pogo" was reborn. According to Pogo, he hadn't been away, the human beans had. "We ain't gone nowhere," Pogo said as official swamp spokesman. "No accountin' for where they been."

The new strip was drawn by Neal Sternecky and Larry Doyle, two college buddies, and distributed by the Los Angeles *Times* Syndicate, which encountered some difficulties in these different times. For one thing, of course, a smaller proportion of the population reads daily newspapers these days. For another, "Pogo" never went for the guffaw. He requires patience, moving at the pace of a seeping swamp, not an MTV comedy sound bite. " 'Pogo' 's not a quick-read, fast-laugh kind of strip," said Douglas Mayberry, a syndicate vice president. "He requires a depth of attention to the strip and to current events, so it has no mass audience, but a select one." Then in a kind of commercialized shorthand, he says simply, "In university towns, yes, 'Pogo' 's read. In mining towns, no."

Still, more than 225 newspapers subscribed, including the Waycross *Journal-Herald,* the sole surviving daily newspaper for eleven (!) counties. My, how the *Journal-Herald* subscribed. In fact, nearly the whole town subscribed to the *idea* of "Pogo," and of what this fictional possum and his fans might do for them and Waycross's economy.

First, they appointed a planning committee and then they created a fall festival, which made Waycross a destination at a focused time. The newspaper and radio station, which certainly have a stake in the area's economy, got on the team, providing the kind of boosterish positive coverage that gave everyone the signal: no nay-saying on "Pogo" and PogoFest. Jim Hart, the executive director of the Waycross/Ware County Chamber of Commerce, even got a little rubber stamp made. "I Go Pogo," it said right around the fuzzy little critter's head. And he stamped his outgoing letters right next to the signature.

Naturally, the committee contacted the two new cartoonists and the L.A. *Times* people and Selby Kelly, who lived out in Oregon but had kept many of her husband's artifacts and was delighted for the attention and honor. Waycross organized a little museum with a rolltop desk, a captain's chair, an easel, pens, brushes, and ink bottles, and Walt Kelly's portable typewriter, all arranged to resemble the Manhattan apartment where he worked for a quarter-century. They had his coatrack, his cane and spectacles, pipes and ashtrays, even a copy of the old Pogo Songbook. It was all warmly approved by Mrs. Kelly, neatly arranged by others, and safely preserved in a town that Walt Kelly had visited possibly twice in his life, near a swamp that Walt Kelly had picked from an atlas by chance. This was a gift of good fortune that civic leaders were not going to let slip away.

Steve Thompson, who runs the international Pogo Fan Club from his home up in Richfield, Minnesota, provided much helpful information and statistics. The club was growing again, with members exchanging information on where best to acquire much-coveted "Pogo" memorabilia.

The new cartoonists contributed fresh drawings of Pogo to help promote the three-day October festival, which was combined with the high school's homecoming. On the eve of PogoFest, the cartoon creature appeared on the *Journal-Herald's* front page in a three-column colored drawing, tipping his hat in welcome. The paper even had a weather box: "Pogo has spoken to the weatherman and all looks good." The newspaper published a large schedule box every day, along with glowing accounts of the previous day's activities, and cable channel 42 broadcast hourly updates of the festival all weekend.

Friday, all day there were historic Waycross walking and bus tours, a film festival on Pogo, a square-dancing exhibition by the Pogo Squares over on Pendleton Street, a Pogo Homecoming Parade, and a railroad handcar race on the railroad tracks. (The Ware County Prison SWAT team turned in the fastest practice times, but the Mighty Fine Boys, a group of friends, actually won the competition.)

Saturday had a car and truck show out at the Hatcher Point Mall, a twenty-mile bike race, a children's Pogo coloring contest, a Pogo item trade show, a bluegrass music jam at Folks Park (BYOLC—Bring Your Own Lawn Chairs), more walking and bus tours and square dancing, and five-kilometer footrace, baby Olympics at Phoenix Park, an exhibition of rap music, a dog show, a remote-controlled airplane air show, capped by dedication of the Walt Kelly Shrine at the Heritage Center, and a show by Billy Joe Royal and the Forrester Sisters at Memorial Stadium.

CSX Transportation, the railroad freight line, donated a diesel engine and four old railcars to provide free rides for everyone on a real train, not one of those toy ones that comes with the carnival that makes the annual circuit like some nineteenth-century Supreme Court justice. Although it is a town that existed because of railroads, virtually no one under middle-age had ever seen a real passenger train, let alone ride one for six miles into the country and back. So the excursions were a very popular novelty. A similar Okefenokee homecoming was later organized around Pogo and July Fourth, which had the traditional fireworks display, which drew the traditional crowds in short sleeves, who uttered the traditional "oohhs" and "aahhs" at the colorful clumps of gunpowder disintegrating in the night sky. As usual, the crowd also had the traditional gaggle of startled babies who vocally disagreed with the joy of these explosions. Then everyone went home all hot and sweaty but feeling good and reassured that life was going on according to a familiar schedule, and the community had shared this feeling— even if a good number of these people thought this Pogo creature was a new commercial creation of some store or restaurant.

Whether it had to do with cartoon characters or mythical local heroes or foods or Indians or a prominent local ethnic heritage, these kinds of festivals were emerging in countless communities along U.S. 1 and many other highways. They were excuses for a weekend of fun, an opportunity for commercialism and special sales, and a point of civic pride that certainly got the organizing committee energized every year, judging by the developing competition for and jealousy about who was picked as committee chairperson. All this might also be turned to lasting civic profit through tourism, the nonpolluting industry that seemed to be growing all over, unlike so many others.

Some of the thirty-five miles of U.S. 1 from Waycross to Florida is divided highway, sometimes with the northbound side completely out of sight behind lush pines that enjoy the temperate climate as much as do the wintertime travelers slowly doffing their winter wear as they get farther south. The community of Racepond was established in the 1830s, when encamped soldiers raced their horses around a pond in between patrols into the swamp seeking Indians who were dodging the enforced trek to reservations in Oklahoma. The countryside is open again with vast grassy fields and broad shoulders, giving a sense of warm expanse. There is an occasional convenience store along the way, usually with one car, the sales clerk's, parked near the door. She's almost always a mother working part-time. And she's almost always replacing the cash-register tape. She takes the time and effort to greet each customer, many of whom are regulars; these are the rural outposts of commercialism where customers and clerks see each other almost daily, or every Tuesday. They smile and call one another by their first names and ask after one another's health, maybe their kids, and likely their thoughts on the weather, which are generally safe subjects in an age of unpredictable strangers. However, they know virtually nothing else about one another, except perhaps if the clerk happens to note how the customer always takes his coffee and likes to buy two chicken wings for lunch from the limited selection of dried poultry parts perpetually baking beneath the lights in the showcase.

Every such store has a modest souvenir selection where the hat pins tend toward the patriotic. The baseball hats concern drinking: "I don't have a drinking problem. I drink. I fall down. No problem." The postcards look colorized and feature unfamiliar sites except for the over-sized local fish who is so big, he sits on the back of a flatbed truck. The soft drinks are always "ice cold" unless they just restocked the cooler, which always seems to happen just before my arrival. The potato-chip selection is always abundant. The hot dogs look leathery and, one suspects, have been slowly revolving there since this trip began. No one beyond the age of five buys chocolate to eat beyond the air conditioning. The newspapers are yesterday's and from Florida. The *Time* magazine is a week old. The girlie magazines are behind the counter or else covered decently by a brown paper wrapper and displayed beyond the browsing range of local youngsters. But the *Soap Opera Digest* and *Guns and Ammo* are in ample supply. *The Atlantic* and *Harper's* must be sold out.

Then comes Folkston, last outpost in Georgia, where someone seems to be a fan of Charles Dickens, who also traveled this highway but not in Florida. On the right stands the old Tiny Kim Motor Lodge. Like the three other motels in town, it is owned by an outsider, which means one of two things, depending on your degree of optimism: either the outsider knows something the locals don't or he doesn't know what the locals know, mainly, the past economic history of U.S. 1 businesses thereabouts.

The Tiny Kim is over thirty years old, a former Howard Johnson's Motor Lodge with fifty units. It is owned by Seoung Kim, an immigrant investor who lives up near Chicago. The Tiny Kim is managed by Irving Kidd. "We still get some tourists," he says, "despite the Interstate." When it opened, the bottom fell out of Folkston's tourist business, which explains why there are so many motels for sale and motel hulks abandoned in the thirty-five miles between Folkston and Jacksonville, used restaurants, too. One empty restaurant looks fairly normal until passersby notice the front windows are gone and a tree is flourishing inside the former dining room.

In recent years, Americans have been taking less lengthy excursions and more minivacations, a few days here and there, a long weekend. Nowadays, they average five-, three- or four-night trips per year of more than one hundred miles from their home. The Tiny Kim's main tourist clientele are retired couples, opting for Route 1's familiar settings, its slower pace, its local flavor. They drift through on the way south in the fall and back again in the spring enroute north, like birds at the refuge. They do not seek an adjacent cocktail lounge with Live! music spewing out the door while, inside, salesmen sit and sip and talk loudly over the synthetic electronic sounds that no one is dancing to. The swimming pool is pretty nippy at those times of the year, but right next door is the Dinkins Restaurant, with a new sign advertising a business policy as old as the stagecoach days: BUSES WELCOME-DRIVERS FREE.

For thirty bucks, they get a clean room with cable TV and a phone; it's ninety dollars a week for some efficiencies. The Tiny Kim is very seldom full nowadays, even with the Trident submarine base not too far away. Irving is happy being 50 percent, sometimes 60, more likely 40 percent full. A few more travelers seem to be opting for the slow local lanes. But he sees very few families come and go; they're passing overhead using up Dad's frequent-flyer miles to visit Walt Disney World. The motel gets some traveling salesmen on per diem, probably a few affairs, too. Who keeps track in these times?

Irving, a former part-time night auditor for the motel, is not into Folkston motel management for the long term. Originally in farming, he left Virginia over twenty years ago. "All my kids are out now," he says. None stayed in little old Folkston. Two are up in Tennessee, one in northern Georgia. Irving'll probably move on at some point, too.

Right now, even without a computerized national room-reservations network to steer travelers to the Tiny Kim and guarantee their arrival with prechecked credit-card numbers, Irving is positioning the motel for the immediate future; he's printed up bundles of motel brochures, which he puts in racks at the rest areas on inter-states for a hundred miles around. Without billboards or the money for radio ads, that's the only way he can catch today's traveler, when they stop going seventy miles an hour to empty their bladders and look at the large plastic-covered map to try to figure out where they are on this cement corridor that bypasses anything local.

There is some hopeful talk about widening U.S. 1 to four lanes all the way down from Waycross past the swamp and Racepond to the Florida border on the south side of Folkston. Irving has learned not to waste too much energy thinking about lines of cars that someday might be lured off Interstate 75 to the west or Interstate 95 to the east and attracted to wander down a wide but local thoroughfare into Florida. Instead, he is anticipating the immediate: the road-construction crews. "They'll come," he said with some eagerness, "and they'll stay five nights a week."

When they come, if they come, they will find in Folkston that the Dairy Queen is OPEN LATE AFTER HOME GAMES ON FRIDAYS. The pavement is cracked. The Plaza Lodge is for sale. The signs at Clark's Amoco say MECHANIC ON DUTY and JESUS IS LORD. But the place is out of business, anyway.

Welcome to the Sunshine State, Where the Mildew Never Sets

FLORIDA'S BACK DOOR HAS a lot of peeling paint. Much is built of cement blocks, and the mortared cracks show the telltale dark drippings of mildew that is fueled by the humidity, which Florida didn't invent but has perfected. Florida is a big place, a single state that is half the size of Italy, with more coastline than California, Oregon, and Washington state put together. U.S. 1 spends the last six hundred miles of its journey south in Florida, entering the state by crossing the St. Marys River.

The highway there is divided by a grassy median and lined by many former businesses—burned, bypassed, bulldozed. Still open is Ridley's Campground, which also buys copper, cans, and old radiators and sells quilts and shell dolls, figuring if you try to make enough ends meet, maybe some of them really will. The state truck weighing station is built in the median to handle traffic both ways and save on manpower costs.

In an effort to survive, many things are converted from their original use. A gas station has become Frank's Antiques. A tractor trailer is hauled into a field, lettered, and left to become a billboard. Terstone's old motor-court cabins are turned into mini-apartments. Besides the pines, the power poles, and the grasses, which locals used to sell as filler for commercial brooms, what seems to grow best in this climate are signs for Ripley's Believe It or Not in St. Augustine.

Callahan was not Walter Swanson's first choice to locate his veterinary practice. Callahan would not exist without Jacksonville. For generations, Callahan's land produced the vegetables, chickens, lumber, and some of the milk that fed and housed residents of the coastal city. Now without much industry of its own, Callahan provides the workers for the city, and not the most affluent ones, either.

A native of Virginia, Walter Swanson fled farther south for the same reason that millions of others have, dislike of the North's long winters and overcasts. He had enough of that in his youth and out at veterinary school at Ohio State University. There, he was a classmate of Larry Berkwitt, the suburban vet back up Route 1 in Darien, Connecticut, who had lost track of his college friend.

Both men wanted to treat small animals and both wanted their own business. So Dr. Swanson shopped the Jacksonville area until he found a community without its own vet. It was not easy to get started. The reason there was no thriving veterinary business was because most people with animals either didn't pay much attention to their animals' health or tried to doctor them themselves, buying the medicine at a feed store, reading the directions, and administering the innoculations. If it worked, great. They had saved a house call or office visit. If it didn't, too bad: God's will.

Some locals still come to pick Dr. Swanson's experienced medical mind and then go do it themselves and possibly return when they screw up and the animal is near death, which means they can blame him. Or the doctor sees them watching him more closely than a paying spectator might, more like a student.

But even just in his fifteen years, he's noticed a significant change. That's because Florida is one of the nation's fastest-growing states. The newcomers went from far north to far south, as far south as they could get in the sunshine state. They bought and built and bought and built and bought and built. And like a warm bottle, the state is filling up. Now even Florida's rugged rural areas in the north near Georgia are finding housing developments emerging, sewer plants overloaded, roads jammed at times, landfills filling.

With their cars and their unused down jackets, these newcomers also bring different values. As Americans moved into the old cities and the suburban cities in their economic orbits, they left behind much of the attitude that animals were mere sources of income. The animals became more companions and guards in a threatening urban environment that could often seem so hostile to some that they took to wearing Walkman earphones as insulating protection from any unwanted sidewalk involvements. This is not a new attitude in American urban history; despite Americans' fascination with and allegiance to countless groups and even groups of groups, despite their attraction to the unnatural orderliness of ersatz main streets in mall after mall, Americans often seem to remain today as they were described 150 years ago by that prescient foreign observer Alexis de Tocqueville, "locked in the solitude of their own hearts."

When it comes to animals, the humans can pamper and fuss and confidently count on them for blind friendship and protection and even safely confide in them. As one result, there are now more dogs in the United States than there are men over eighteen. As another, medical research, cosmetics, and other industries that had used animals found themselves the object of emotional attacks by so-called animal-rights advocates. In some cases, such as up U.S. 1 in Norwalk, Connecticut, these attacks could turn violent. In others, such as at the Agricultural Research Center on the same highway, infiltrators kidnapped the animals to "liberate" them.

In Callahan, Dr. Swanson can see noticeable changes among his new clients. They are used to having pets and treating them like family members, which means health maintenance through preventive medicine. "These folks don't wait until the animal is near death's door," says the doctor. About three-quarters of his work concerns dogs now, the rest cats, rabbits, hamsters, and the like. "I don't mess with alligators," he adds, only half-joking.

He has office hours six days a week right on U.S. 1, seeing maybe eleven appointments and a few walk-ins. Most of his customers pay with checks on Jacksonville banks. He lives by himself above his office. "So, I'm here for emergencies," he says.

Unless it's lunchtime, when everyone knows they can find the doctor and many other regulars just down the street at Ann's Diner. Ann is Ann Bradley, Claude's wife. They're forty-two years out of Ontario, working their way south from Canada to Vermont to Connecticut to Callahan. Her business has changed considerably in the years since the Interstate went in, she said. Now, 85 percent of her customers are local. They know about her famous chicken and beef lunches but couldn't care less if Al Capone was among the famous, infamous, and inconsequential customers who sat at the diner's tables over the years. "Sure, it's gettin' tougher," she said.

It is not hard to tell that U.S. 1, for the first time since Connecticut and New York, is back near the ocean; radio station disc jockeys, for whom the word *blabbermouth* was invented, are now frequently announcing the time for the next high tide, which might be helpful should any moronic fishermen have sailed without their own tidal tables or if you happen to be driving a car on Daytona Beach. The radio stations also provide the latest "instant up-to-the-minute accu-traffic announcements," which are extremely informative unless the details happen to concern where you are driving at that very moment, in which case the accu-details are most likely inaccu.

Florida radio announcers are handicapped, however. They must say things like, "Tonight, temperatures plunging as low as the mid-forties." And they must wait for irregular hurricanes to issue doomsday forecasts. Whereas their hardworking colleagues back up north on U.S. 1 need only wait for winter to verbally wring their hands—

"Freezing rain, sleet, and possible accumulations of *three* inches of snow overnight. So if you don't have to go out, better stay inside today." Tomorrow, too, and the following week as well, just to be safe. This, not coincidentally, will keep your ears near the speaker, ready to receive every air-filling broadcast sports cliché ever invented for the Eastern Seaboard: Hey, how about those Sox? How about those Mets? How about them Phils? How about those Birds? How about those 'Skins? How about that Magic? How about the Heat? How about those Dolphins? How about, who cares?

One of the nice things about Jacksonville, besides being one of the blessedly few American cities intentionally named for a president, is the distance it has managed to maintain—until now—from the usual local fawning over a professional sports franchise. This fawning is truly an obnoxious blight, unless it involves the Cleveland Browns, which is understandable, though too far off Route 1 to explore here. Jacksonville's sports passions are divided between the Gators of the University of Florida in Gainesville and the Seminoles of Florida State University in Tallahassee, which remains to this day the only Confederate state capital east of the Mississippi River *not* captured by Sherman or some other renegade in blue. This record seems likely to stand.

Several Saturdays every autumn, thousands of Floridians like John and Gena Delaney trek up and down their main highways to pledge athletic allegiance to their college team. The Delaneys (he's an attorney, she's a nurse) met in Gainesville. Their team is the Gators (despite his college education, John sometimes also roots for an NFL team in Cincinnati). The Delaneys get Gator season football tickets and sitters for the kids. They perhaps don a piece or two of clothing in the proper colors. They jump in the family car, and they drive to Gainesville to have a wonderful time with friends who are also attaching themselves to the successful program and the slightly faded memories of fall Saturdays in their youth.

These are not long trips. But no trip is ever too short really for John to indulge himself in one of American maledom's most secret and satisfying travel pastimes, The Silent Mileage Competition. Just about every male does it. Very few ever think to talk about it; it seems so unremarkable, natural, and necessary, like breathing. There is obvious competition at work, on the office softball team, or perhaps the couples' monthly volleyball games, which the wives go to play and the men go to win.

So when a male finds himself in a car and no one to compete with (what's he going to do, race the kids in the backseat or challenge his bedmate to dueling carburetors?), he creates a mental competition. Him against the clock, with the odometer as referee. It can even happen late at night. Suddenly, the darkness is shattered as John turns on the vehicle's ceiling light. "Are we home, Daddy?" asks a sleepy voice from the back. "Honey, what's wrong?" says an instantly alert woman by his side.

John is doing quick mental calculations. "Yes, oh, yes," he exults, and turns to make the proud announcement. "You know, we just did seventy-four miles in seventy-three minutes!"

Gena, speaking for every other uncomprehending female on the continent, looks at her husband, waiting for the rest of the revelation, and, with her mouth open, says absolutely nothing.

John, eager for the adulation he is certain will be forthcoming any minute now, has nothing more to say. Seventy-four miles in seventy-three minutes! That pretty much says it all.

Gena nods, thinking back quickly over what her husband ate for dinner a few hours earlier.

John nods, too, awarding himself an imaginary high-five. And he quickly notes the mileage now to begin the new competition before turning out the light on another episode of The Gender Gap.

Another nice thing about Jacksonville is that U.S. 1 is so clearly marked all the way through town. Even when it briefly joins I-95 before ducking down local lanes once again past houses past their prime, there is no doubt where 1 runs. Along a row of pawnshops, jewelry stores, vacant lots, liquor stores, and the Gator Auto Insurance agency, there is no doubt. And on through the tidy downtown, where fewer strollers have been seen ever since air conditioning reigned supreme, before crossing the St. Johns River and heading out of town as the four-lane Phillips Highway, which would be an expressway except for the stoplights dropped in to help the nation boost its gasoline consumption.

Ponce de León discovered Florida in 1513, which was news to the Indians who had lived there for quite a few years and had not advertised any attractions for wandering Spaniards who missed the Bahamas. Mr. de León was seeking the fabled Fountain of Youth theme park, and maybe a little gold and silver, if any was lying around. He found an Indian arrow instead. No one remembers quite why he dubbed that humid peninsula Florida; *florida* means "flowery" in Spanish, or maybe it was for Easter, *pascua florida,* the alleged day Ponce de León landed near what would become St. Augustine. According to St. Augustine's proud visitors' bureau, this landing and subsequent settlers' arrivals were important. "To this day," the bureau reports, "the city remains the oldest permanent European settlement in the continental United States." As if this requires any physical effort to keep other cities from sneaking in earlier in history. See, the Vikings' etched stones back up in New England can't count because they're no longer inhabited by Vikings and the Pilgrims and their brand of limited religious tolerance were still fifty-five years from Plymouth Rock.

One thing's for sure, Florida could never be called Pacifica—too much fighting all the time. After Ponce de León came Hernando de Soto, who, in 1539, began his ruthless trek from there across North America, but still got a car named for him by automakers who don't read history books. In 1565, the Spanish were bothered by a settlement of Huguenots, who seemed to bother all kinds of people, which is why they had to split up. So the Spanish wiped out the first community on Jacksonville's site. Then there were the British to contend with. They got Florida in a swap for Cuba, named the Jacksonville area Cowford, and left after twenty years. Then there were the pirates who were after all those galleons that always waited for the hurricane season to waddle up the Gulf Stream. And then around 1812, the Americans wanted Florida for extra protection. But they already had one war going with Britain, so in 1819, they bought the place. That was not the last Florida real estate deal, though it was one that sealed the Indians' fate; after some violent altercations, they were sent to an inland retirement place called Oklahoma. And Florida, the staunchly pro-slave state, was left to moulder in the tropical backwaters even during much of the Civil War. Then, in the 1880s, came the arrival of Henry M. Flagler, who was not retiring.

Mr. Flagler was a native of upstate New York; he worked in an Ohio general store and then made and lost a fortune selling salt in Michigan. He returned to Cleveland, where his business acumen and ruthlessness led him to set up his desk in the same room with someone named John D. Rockefeller. They were figuring out what to do with this gooey black stuff that could be refined into kerosene for light but left this other volatile liquid called gasoline. Although Flagler's part in founding and building and building Standard Oil and destroying the competition is less well known, by all accounts he did play a crucial role. And then following doctor's orders to recover from a stubborn liver ailment, Mr. Flagler passed the winter of 1883–1884 in St. Augustine. Like George Washington and the pewter spoon episode in Connecticut, Flagler was put off by Florida's poor accommodations and service.

There might have been other rich men with vision in those modest rooms by the sea, but none with the personal fortune and the ambition of Henry Flagler. And he set out to correct the local situation, combining some savvy with a lot of money, some good luck, a warm climate, and an American penchant for greed. The result is today's fourth-most-populous state, a tax and weather haven for retired couples, vacationing families, foreign tourists, cruise ships, and a space industry. There are about 13 million residents of Florida; every year, more than three times that many people come to visit, rent a car and a modern hotel room, and pour money into an economy based in large part on Mickey Mouse. And they're not just Americans, either. On any given winter day, 1 million Canadians are wandering around the Sunshine State listening to special radio newscasts of Canadian news and weather, which make them feel even better than the warm sand.

A cartoon character with big ears and a squeaky voice is not what first comes to mind upon viewing the dour, mustachioed visage of Mr. Flagler, who went through three wives with this Florida thing of his. The first Mrs. F. died. The second one went incurably crazy in Florida, prompting the state legislature to pass a measure in 1901, making incurable craziness sufficient grounds for divorce. What a coincidence! On August 14, 1901, Mr. Flagler obtained a divorce from Mrs. Flagler. And ten days later he married wife number three. It must have been love at first sight. She got all the houses and money when he died on May 20, 1913.

But in the meantime, Henry Flagler changed the face and the economy of the southeastern corner of the United States. He had an idea that if Americans could get to Florida and if they found luxury hotels and first-class service even in tropical heat, then there might be some money to be made. He would create an American Riviera. He began in 1886, buying the Jacksonville, St. Augustine & Halifax River Railroad and then, in a pattern familiar to Standard Oil's shareholders, he bought up other little local railroads. He assembled them into the Florida East Coast Railway and began extending its lines steadily down the coast, first past Daytona, then to Palm Beach in 1894, and finally to a jungle clearing called Miami in 1896.

Along the tracks, he built a platoon of palatial hotels beginning with the Ponce de León and Alcazar in St. Augustine, the Ormond at Ormond, the Royal Poinciana and the Breakers at Palm Beach, and the Royal Palm in Miami, where he had the harbor dredged to permit ocean shipping and cruise-ship traffic to Nassau, where he built two more luxury hotels. To fill up the freight cars of his railroad, he helped encourage agriculture; citrus fruits seemed especially suited for the coastal climate. Then he stretched his tracks fifty miles through the muck of the Everglades. Then he took on the real challenge: building a railroad across 106 miles of tropical waters and coral islands all the way to Key West to connect with the ships to Cuba. It only took seven years. The first train ran over the ocean in January 22, 1912. By that day, Flagler had invested in Florida nearly $50 million in real money, pre-income tax. That does not count the millions more in anonymous contributions he made to hospitals, schools, and libraries as part of an old-fashioned social contract among American multimillionaires such as Carnegie, Ford, and Flagler's Standard Oil partner: Once you get so much money that you need a staff to count it, you give some of it away, anonymously if you want extra attention.

Other than that, though, Henry Morrison Flagler, the unschooled son of a Presbyterian preacher who never made more than four hundred dollars in a year, had no impact whatsoever on the future of La Florida.

Flagler made one mistake: He underestimated his own impact. His goal, as outlined to a friend, was simple. "It occurred to me very strongly," Flagler said, "that someone with sufficient means ought to provide accommodations for that class of people who are not sick, but who come here to enjoy the climate, have plenty of money, but could find no satisfactory way of spending it." Thus, as must be obvious by now, the entire resulting development of Florida was really an act of charity to help relieve vacationers of the burden of possessing too much money. In that sense, Flagler's hotels, Disney's endless theme parks, movie-studio tours, and countless alligator farms and souvenir shops have done an admirable job.

One problem was, however, that for several decades until Walt Disney World and its imitators made tourism the number-one industry in their neighborhood, Flagler's developing attractions in southern Florida undercut his own operations in the north in Florida. Who would stay in St. Augustine above Florida's frost line when by staying on the train or highway for another 250 miles they could suffer through the powerful warmth and aura of Palm Beach? Even today, promoting themselves as "First on Florida's First Coast," St. Augustine and St. Johns County attract fewer than 2 million visitors a year, an avalanche by Waycross standards but barely enough to keep the area in the top eight of Florida's most popular tourist destinations. As one result, to survive in its Spanish Renaissance splendor, the luxurious 245-room Ponce de León Hotel was shuttered and converted into Flagler College, at a renovation price about ten times that of its initial construction. A couple of blocks away, the old railroad still operates, although the number of piggyback trucks perched on railcars far exceeds the number of people perched on passenger car seats.

After the king of Spain took his capital gains in 1819 and fled, a handful of real estate companies tried to sell Florida real estate sight unseen. They had modest success creating pre–Civil War leisure-time communities due to the unfortunate impact of hurricanes, yellow fever, and Seminoles waving hatchets, which will cut down on contract land sales every time.

"Keep your head above water," Mr. Flagler opined, "and bet on the growth of your country." So, even without the Fountain of Youth, Florida's land boom exploded in the 1920s, thanks in part to the hired-out eloquence of William Jennings Bryan, who redirected his golden tongue from denouncing Darwinism to extolling the guaranteed investment virtues of developments such as Coral Gables just south of Miami. Palm Beach was touted as an exclusive community, as was Boca Raton, whose name sounds fancy until it is translated. The first day's sales of lots in Mouse's Mouth, Florida, topped $2 million in 1925 dollars. Many of the lots were above water. Sixty-five years later, the Boca Raton condo market would prove irresistible to, among others, George and Marianne Vecsey, who was alone at the motel on Route 1 in Boca Raton the night before the closing.

Palm Beach County is consistently one of the nation's fastest-growing counties, adding 1,700 new housing units and two thousand new residents every month. Statewide in the last decade, Florida's population grew by 3 million, thirty-five newcomers every hour around the clock every day for ten years. Of these new residents, fully 90 percent are retired. People over sixty-five make up 18 percent of Florida's 12.8 million residents, the highest rate of any state.

This cast of newcomers is a welcome boon to any state because the retired bring in the accumulated savings of a lifetime, including the substantial proceeds from the sales of their homes up north, while they make very few demands on the social infrastructure, at least while they are healthy. As one result, only 10 percent of Florida's workers labor in factories, where layoffs come most often in difficult times. That's barely half the national rate. Another 30 percent, highest among the top ten states, work in the hardy service sector. This has helped protect Florida from the recessions that regularly pain other states.

The sales tax structure is heavily geared to hit visitors, which helps keep the total tax burden low (fourth-lowest in the country), which helps feed the cycle of population growth and so on. The tax burden attracts industry, too, including, as already noted, nonpolluting job-creators such as the American Automobile Association, which moved down Route 1 from Virginia to settle slightly inland near Orlando. NASCAR has its national headquarters just two blocks off Route 1 in Daytona Beach, which with adjacent Ormond has long been an international symbol of sand and speed. Ormond was founded in 1873 by Connecticut's Corbin Lock Company, which wanted a warm-weather resort for its tubercular employees. In 1937, with his golf game still not perfected, John D. Rockefeller died there at the eccentric age of ninety-seven, until the very end handing out dimes to encourage thrift among passersby. Daytona was founded in 1871 by another Ohioan, Mathias Day.

Where there is a relatively flat surface and a motor or wind or a horse, some kind of racing tends to erupt among Americans. By 1902, the top speed on the twenty-three miles of hardpacked Daytona area beaches was a mind-boggling fifty-seven miles an hour. A Stanley Steamer got up to 128 in January 1906, which just hap-

pened to be the height of the tourist season. By 1935, when the land-speed record-seeking business was forced to move to Utah for adequate running room, Sir Malcolm Campbell took his rocket-powered car up to 276 miles an hour. The precise reason for doing this is not recorded. Today, the beach speed limit is ten.

But when southern stock-car racers (the true kind, understand, as opposed to some Yankee pretenders) began to compete on a nearby oval track, the crowds of spectators grew so large that a bigger track was necessary. Today, the five-hundred-mile race at the 2.5-mile Daytona International Speedway kicks off the twenty-nine-race main NASCAR racing calendar every February. Racing has become such big business that NASCAR sanctions hundreds of other races in eleven different auto divisions at 115 different tracks across the country.

In one week of February, just the Daytona track attracts 350,000 race fans. Combine these folks with the spring-break college crowd and assorted other families plus, of course, a lost author working his way down the hemisphere from 7-Eleven to 7-Eleven and that's a pretty good living for the Daytona area's 16,000 hotel rooms, suites, and apartments. The young people appear every spring after exams, making their way south on the highways like a wave of large insects, shedding their sweaters with the rising temperatures and consuming all the pizza in their path. The uniform of the day is swimsuits, which the females always seem unaware they are about to spill out of. The males, however, seem quite aware. Whenever I see such collections of sun-seeking hormones, I am reminded of the elderly retiree encountered standing across the street watching such beach doings some years ago. The only thing wrong with today's younger generation, the woman said, was that she was not part of it.

U.S. 1 is paralleled along much of Florida's eastern coast by A1A, which is closer to the water and the bikini views. Old 1 does not always pass through the classiest neighborhoods. There is the Capri Movie Theater, which, according to the sign, features BOOKS AND NOVELTIES— ADULTS ONLY. Kitty-corner is Where the Girls Are—

WE'RE STILL TOPS IN DAYTONA, A WORLD CLASS SHOWGIRL REVIEW—although it does not have any windows. Nearby, is the Sans Souci Motel ($15 SGL). To cut into two different markets from the same building, Lea's Pest Control shares quarters with a tanning salon, in case nature's rays produce an insufficient amount of skin cancer.

And then there's Angelo's Central Florida Tattoo Studio (not a parlor, get it straight). Joe Collins works there, twelve to eight, seven days a week. Business is really good, he says, half of it provided by women. "Guys go for the bigger tattoos," Joe says, the younger ones preferring your Bart Simpson tattoo or your college fraternity letters. The older guys go more for dragons, tigers, and wizards on their arms or backs.

The ladies tend to choose more butterflies, little hearts, and tiny roses, which they have permanently inscribed on their hips or their chests, just hidden by you know what, or on their buttocks, where only a certain someone or several someones will ever see it. "See," said Joe, "it's more sensual in a discrete area." He reports that the women, often spurred on by their boyfriend, do not seem embarrassed to take their pants down and have a strange though very nice man stick a vibrating needle into their skin a few thousand times. "By the time they get in here," he says, "they're not that shy."

Times have changed in this age of AIDS. Joe wears rubber gloves now and he says reputable shops like his put on new needles for every customer. He's even had one customer return from Switzerland three years in a row to get new tattoos or complicated additions to old ones.

A simple ten- to fifteen-minute butterfly will run you about thirty-five dollars, Joe says. A decent dragon might take two and a half hours of needling and set you back three hundred dollars or more—no extra charge for the continual burning sensation.

As to why people choose to pay such sums to be tortured, Joe says he has no idea and doesn't really care. However, a *New York Times* feature writer recently found a doctor, described as an expert on the psychodynamics of tattoos, who explained:

> The tattoo can be viewed as a psychic crutch aimed to repair a crippled self-image, inspire hope, keep noxious emotions at bay, and reduce the discrepancy between the individual and his aspirations.
>
> For many young individuals, the tattoo is a pictorial quest for self-definition, easing one's inadequacy and isolation. It offers the tangible promise of a final identity, the clarified picture of a diffused ego.

On the other hand, on or off Route 1, a tattoo could be just a youthful stupidity that won't wash off.

One of the (JUST 3 MILES AHEAD MANNY'S INDIAN MOCCASINS) more interesting aspects about life (MANNY'S DAYTONA T-SHIRTS 2½ MILES) so relatively recently emerging from pioneer days (MANNY'S SOUVENIRS SPORTSWEAR DISCOUNT PRICES 1 MILE) is the abundance of (MANNY'S ADULT SWEATSHIRTS FLORIDA DOLLS ¾ MILE) individualistic promoters (MANNY'S SCENIC POSTCARDS 8 for $1, ½ MILE) who still try to make (MANNY'S FLORIDA BEACH TOWELS $3.95, ¼ MILE) a decent living (MANNY'S STRETCH YOUR VACATION DOLLARS ⅛ MILE) serving tourists (MANNY'S FRESH COCONUTS 69 CENTS) wending their way (MANNY'S BEACH MATS JUST AHEAD $1.49) along the historic (MANNY'S FREE VISITOR INFORMATION JUST AHEAD) local roadways of (MANNY'S FREE LIVE ALLIGATOR EXHIBIT SLOW JUST AHEAD) eastern Florida (MANNY'S TREASURES OF THE SEA HERE) where so many have traveled before (THIS IS MANNY'S!).

So it comes as something of a shock to learn that Manny is no longer at Manny's. Manny is dead (WELCOME! MANNY'S GRAVE JUST AHEAD). However, while Manny attends to the great curio counter in the sky, travelers on U.S. 1 wearing the traditional vacation uniform of sunglasses, Daytona T-shirts, and wash-and-wear shorts that do not fully cover the patterned red calf marks from the vinyl-covered car seats will be relieved to learn that Manny's does survive. Any tourist who finds her supply of Orange Blossom Cream Perfume running low can still stock up at Manny's, which carries mail-home packs of the secret-formula substance, as well. If you've misplaced your last rubber alligator, Manny's staff has plenty more. Same for the rubber Florida beach sandals from Thailand, the plastic alligator heads on sticks that actually work for an hour or two in the backseat, the plastic back-scratchers shaped like a dwarf's hand, and the seashells, which once resided in the sea for free. Did you lose your pink rubber coin purse with the white Florida flamingo on it? Break your last Florida ashtray? Need another dried starfish to complete a collection? Manny's still got 'em.

Of course, none of the 150,000 people who still stop at Manny's every year, not knowing that he has moved on to peddle plastic angel wings to the long line before St. Peter, none of these living people ever actually buys any of this tacky stuff. I mean, c'mon! They just want to stretch their legs, right, while the temperature of their dashboard nears that of the sun. I almost didn't park by the cement dinosaurs. I almost got to sleep for a month without the word *Manny's* leaping before my closed eyes. By the time I got to Manny's, he seemed like an old friend and I felt right at home, laughing at the array of junk in a superior kind of way. It was better even than the Route 1 Flea Mkt. back in Jersey. But, hey, who would ever pay $2.98 for a straw sun hat with a green palm tree and the word *Florida* embroidered in red on the brim? Or Florida pens with sea horses floating, entombed, in the plastic? And whatever would I do with a lensless pair of oversized plastic glasses with a rubber nose and mustache attached? Although, you know, $1.59 is not a bad price. And it was the last one.

Florida's U.S. 1 does have other attractions (YOU'VE JUST PASSED MANNY'S—GO BACK!). And an endless supply of roadside exclamation marks!! There's Granny's 24-hour Catfish Restaurant, Taco Ted's, and the ABC Liquor Store, which is open at 8:00 A.M. for your early-morning convenience. The Santa Maria and Sandman motels are both out of business. But the Frontenac Flea Mkt remains—LARGEST FLEA MKT OF ITS KIND ON FLORIDA'S EAST COAST! (How many different kinds of flea mkts can there be?)

But my favorite roadside attraction of yesteryear was the "All-U-Can-Drink" orange-juice phenomenon, which George Harvey remembers not so fondly. "I didn't want to do it," George recalls. "But back in the fifties, there were twelve fruit stands along here and they started offering All-U-Can-Drink juice for ten cents. So I did, too. Until one day this fellow comes in, puts down his dime, and starts a-drinking. Six glasses. Seven glasses. Now in those days, we did all the squeezing by hand, mind you. Ten. Eleven. Along about the fourteenth or fifteenth glass, I said, 'Wait a minute, young fellow. I've got another gift for you.' I went out front, pulled up that 'All-U-Can-Drink' sign, and gave it to him. Haven't had another one since."

Farming was the last thing on George's mind growing up seventy-seven years ago on Cleveland's West Eighty-third Street. At fourteen, he was sent down to Florida to help his aunt and uncle, Roy and Blanche Harvey. They hadn't intended to stay in the Indian River area, either. But Roy was not in the best health and neither was their car, which broke down on Route 1 in 1924. They stayed a while to earn money for repairs. Roy was a mechanic. Blanche waitressed at the Cocoa House Hotel, serving food to the likes of Eddie Cantor and Al Jolson on their way through. The hotel site is a bank now.

One day trying to earn another dollar or two, George's aunt bought a box of grapefruits and a box of oranges. She put them by the roadside on a board between two crates. She sold out. The rest, as they say, is history.

Today, George Harvey, the teenage hitchhiker who used to caddy for rich golfers near Cleveland, owns 1,100 feet along U.S. 1 plus another three hundred acres over on Merritt Island. He employs upwards of ninety workers at busy times, and, in winter, dispatches several trucks a day up Route 1 that are crammed full of boxes of citrus fruit, which get mailed to customers in northern cities from post offices there because the mail from Rockledge, Florida, is so slow.

George was one of the all-time opponents of building Interstate 95 some twenty years ago. He lost. He knew it would ruin his roadside business. He was wrong. "I thought the Interstate was the end of my life on the local road," George says. However, it seems high-speed auto travelers are looking for an excuse to slow down and stop and don't mind a two-mile detour from Exit 74.

Harvey's is open from 8:00 to 5:30 seven days a week now from mid-October to the end of May. The business generated by the four billboards advertising his exit off I-95 and the ten signs along U.S. 1, the eight WATS lines, and the 300,000 mail-order catalogs he sends out every year produces orders for over 9 million pounds of citrus fruits (eight varieties of oranges and three of grapefruit) and gross annual revenues exceeding $3 million.

With local sales and the general emphasis on healthier eating among the younger area families who seem to be increasing, business is so good that they had to put in a stoplight, of all things. Sales are best on football or Daytona race weekends—one thousand customers or more a day. Some, wearing those little safety beanies, even arrive on bicycles; adult bicycle riders are big orange-juice drinkers—very healthy, you see. George has some customers who bought his oranges forty years ago with their children, now retired and still buying his fruit.

Both George's boys—Larry and Jim—help with the business, as does his wife, Jannetta, who lives with him in their gracious colonial home a 142-second commute away through the sandy groves. The family even uses photographs in the catalogs of George's two towheaded grandsons, smiling, standing on a ladder sucking lush oranges. Many of Harvey Groves' seasonal employees are the wives of the thousands of workers attracted to Cape Canaveral and the area's numerous defense or aeronautical industry plants, women like Jan Sherrington, who can charm nearly anyone with her Scottish accent. Since he feels they represent him, George still personally interviews every one of the store workers, seeking, above all, openness and friendliness, which he finds harder to find these days.

But modern citrus farming can be a tricky business. Handshakes and a man's word seem insufficient to hold some deals together; that takes at least one lawyer and lots of signatures nowadays. With weather and disease and the financial temptations of selling land to housing developments with names like Quail Hollow, only two roadside fruit-stand operations remain in the area.

In that bad frost a couple of years ago, the Harveys lost thirty-seven acres of trees. With each acre holding an average 110 trees and each tree producing upwards of twenty-five bushels, that's a lot of fruit that wasn't. "I hate to take up those good trees after all these years," he says. "They become like family."

While orange trees begin producing fruit by the age of five, they don't reach their peak production for twenty years or so. And some continue productively for ninety years. But first you've got to get the trees; citrus nurseries have a two-year waiting list.

There is, however, no waiting line for fresh-squeezed orange juice, though it now costs forty-five cents per glass.

Maybe it's the rocket motor standing idly in front of the Titusville High School (Home of the Terriers), or the Chamber of Commerce "affectionately" nicknaming the town "Space City, U.S.A." (What exactly is affectionate about those words?) Maybe it's the signs for the Space Camp (one-year waiting list) or the Astronaut Hall of Fame. Or maybe it's the hundreds of thousands of people who regularly gather along the shoulders of Route 1 to gaze toward the east where some immense buildings the size of mountains seem to rise out of the saltwater horizon.

But it doesn't take a rocket scientist to figure out the area's economy has a lot to do with leaving this world at a high rate of speed. Especially when the air begins to rumble, even seven miles from the launchpad. And the night sky lights up. And the thunder grows louder and louder until it buffets the ears. And that man-made machine with people aboard or maybe a satellite or some secret unannounced cargo slowly climbs off that sandy spit up through a horde of startled seabirds who never saw any sea gull that big or noisy before. Then, faster than anyone can say *Fahrvergnugen,* that creature rumbles off to the southeast toward Africa or Mars, whichever comes last.

Incidentally, Canaveral is not a Spanish euphemism for barbaric eating habits. Capo de Canaveral (Cape of the Wild Canes) began showing up on old maps (well, they weren't old in the early sixteenth century, but they are now) after a Spanish slave trader, Francisco Cordillo, stopped by on a recruiting visit. The natives objected to such enforced enlistment and fought off the Spaniards by cutting wild cane stems from local marshes and hardening their trips to become painful arrows.

The rocketship is a further refinement of that idea. Once, when the Germans fired their rockets, everybody on the other side ran and hid. Now that the Germans are on our side, at least for now, most of them long since retired even, people flock to watch these metal behemoths compete against gravity. Back in 1966, a busy day saw five hundred tourists stop by Cape Canaveral for a crude tour. Today, four thousand visitors is a quiet day. Space tourism is such a big business, in fact, that it outgrew even the National Aeronautical and Space Administration, which leases the tourism concession to a private company, TW Recreational Services, Inc. More than 3 million people a year from every state and sixty countries go through Spaceport USA (remember now, this is different than Space City, U.S.A.). The Spaceport alone (actually, it's a very fancy and imaginative visitor's center) employs five hundred people, who pump more than $10 million into the local economy. (The Cape employs 16,500 others.)

Everything about Spaceport is big, very big. Its thirty-eight buses log nearly 900,000 miles a year just showing visitors around the Cape and its famous launchpads (that's more than two round-trips to the moon, in case you're keeping track). The IMAX movie screen is five and a half stories tall and provides even middle-aged journalists with the heart-pounding, wraparound experience of standing next to a launchpad at launch time. The ground-floor souvenir shop sells every imaginable space-related item. Next door in the cafeteria, voracious space tourists whose anatomies may also be vainly fighting the forces of gravity consume 136 tons of french fries every year. Hamburgers nose out hot dogs in the annual space-food consumption race, sixty tons to fifty-seven. Spaceport has thought of everything; they even have dog food and air-conditioned hourly kennels to keep family pets from baking to death in the sun, forgotten, in parked cars.

Because foreign tourists tend to stay longer (two weeks versus four or five days) and spend more, Spaceport, like many Florida attractions, is aggressively traveling the earth to solicit more of them. Foreigners now comprise 15 percent of the state's tourists, a seven-fold increase in five years. And Spaceport is not the only theme attraction on the Space Coast. Within an hour's drive is Disney, Epcot, Universal, and all those other tourist attractions, which is a little-noticed but very dangerous concentration of theme parks. That means that during a nuclear war, a single missile could wipe out a major portion of Florida's entertainment enterprises. Although Parrot Jungle would likely be outside the blast zone and thankfully spared, such a holocaust could take several tourist seasons to repair. However, come to think of it, it would provide the theme for a new park with some potential, including a museum on radioactivity.

Today's Space Coast is a far cry from the mosquito-laden early 1800s when Henry T. Titus set up a general store there. A huge man, Titus stood over six feet tall in his long beard. A self-appointed colonel, gunrunner, and soldier of fortune who fled Trenton, New Jersey, in his youth, Titus had a penchant for going off to join Cuban revolutionaries such as Narciso Lopez. This adventure ended with the untimely capture and sudden expected death of Señor Lopez. But there was also some excitement with the Coast Guard and then some scrapes with John

Brown and the Union Army up in Kansas. However, his reputation kept other troublemakers away and spawned a certain prosperity even among "crackers," hard-bitten cattle drivers named for the sound of their long whips.

Titus fathered many children in many places. This may have caused a hormonal imbalance, because, according to some accounts, he suddenly turned into a regular homebody who liked making shingles and planting shrubs, which he did until his death in 1881. In the colonel's honor and because he had the largest gun in town, the developing community was named Titusville, which a few of today's 40,000 residents know about.

Titusville may have missed its best shot at world-famous glamour because of another rugged pioneer, Louis Coleman. He owned a large chunk of land there called Sand Point, which a real estate agent for Mr. Flagler sought to buy. Louis (he wasn't rich enough to earn a Mr. in this book) put too high a price on his land. So Mr. Flagler took his extravagant resort plans farther south to a place to be called Palm Beach.

When Howard Benedict arrived in 1959, Titusville was a sleepy, pathetic little place of uncounted citrus trees, several billion insects, a few thousand cows, and five thousand humans. Mr. Benedict was a young reporter for the Associated Press, and those were the infant days of the American space program just before Senator John Kennedy of Massachusetts talked about a missile gap he thought he spotted threatening the country's security. Fortunately, the gap disappeared after his election as president. So the country could get on with the business of catching up with the Russian space program and getting to the moon first because there might be something there.

Beginning with the original *Mercury* 7 astronauts, it was Mr. Benedict's prose that vividly brought the drama of all those early epic space flights to readers of newspapers around the world. They had twenty-nine launchpads working in those years, launching many military missiles, too. Of course, it was the manned shots that drew all the crowds of technicians, spectators, newsmen, astronauts, and politicians who were quite quotable and

quite certain that this latest launch, whatever it was, was symbolic of the never-ending American quest for new frontiers to conquer, even if this latest frontier had been traditionally left to the deity. The fact that a fair number of early rockets got about fifteen feet off the ground before exploding only temporarily dampened the profuse political pronouncements of support.

AP stories flashed out of the Cape (up Route 1 in New England, there is only one Cape and it has nothing to do with rockets, but for the rest of the United States of America, the one Cape is in Florida) every few days. The country had a new set of clean-cut heroes with smiling wives and proud parents and names like Shepard, Cooper, Glenn, Schirra, Grissom, Slayton, and Carpenter. There were some pretty good parties thrown by visiting contractors along Cocoa Beach and Cocoa, which got its name when the post office rejected Indian River City as too long to fit in a postmark (hearing this, the boys gathered in the general store sought a new name, and the first place one pair of eyes fell was on a nearby grocery shelf where some Baker's Cocoa was sitting). You might actually see the crew-cut heroes around the Vanguard Motel, a popular gathering spot for the space and news crowds, not least because that's where the Associated Press had its news bureau.

The military paid these pilots all of $12,000 a year. *Life* offered $500,000 for the story of their earthly and space lives. Brevard County (Theodore Washington Brevard was a pre–Civil War state comptroller) might get a million visitors for a dramatic launch like one of the moon shots. Residents along U.S. 1 would charge ten dollars a car to park on their lawn.

Today, of course, rocket firings have lost some of their novelty. U.S. 1 has gone from being the state's main through road to being a through paved backwater to being a main pain for local traffic, the busiest street in the area, handling some 35,000-plus vehicles per day. Somewhat fewer spectators drive along it for the big Space Shuttle shots. There have been and are now so many astronauts that no one beyond the families can really keep track of each, and the Spaceport's remnant bin is full of pins picturing each forgotten team in their white space suits. (You expected American astronauts to wear some other color?) Titusville has blossomed into a regular city with modest homes fetching $700,000 to $800,000.

As they have all over the country, several local newspapers have been bought up by a chain (Gannett) and merged into one. *Florida Today,* the bright, breezy, colorized, no-sweat-to-read newspaper was the prototype for its bright, breezy, colorized, no-sweat-to-read national version, *USA Today.* (So rushed. You see. Readers are. Big rush. No time to read. Much. Unlike old times. Gotta save time. For some reason. Gotta save space 2. No space for periods. Even in USA. Like TV.)

Also changed, the Vanguard Motel has been bulldozed for Alan Shepard Park. The space business has gone through some booms and near busts, but space has an endless supply of frontiers to conquer and, down here, political deals to forge. There is growing concern at the aging Cape Canaveral complex about European competition for the commercial launch business. The military and space contractors have teams based near the Cape full-time now. The causeway out to the Cape is four and sometimes six lanes. John Glenn is a senator now. Gus Grissom is dead.

Howard Benedict is retired from reporting. Today, he is executive director of the Mercury 7 Foundation. The Foundation runs the Astronaut Hall of Fame, which sells rooftop spectating space for ten dollars per person per launch; the Space Camp, which takes two hundred kids a week twenty weeks a year and has a long waiting list; and a college scholarship fund for science and technology students. The six remaining astronauts still get requests for personal appearances, some from as far away as Japan's Space Camp. The men accept most invitations. They donate their fee to their scholarship fund, which is halfway to its goal of a half million dollars. But the space pioneers along U.S. 1 have set some limits on their commercial appearances. "They don't do shopping centers," said Mr. Benedict.

* * *

Of course, progress has its price, which is not so bad if someone or something else pays. Some of Brevard County's wildlife have been unable to adjust to all the growth. Mastodons once roamed there and natives who invented a spear-throwing device and a club with a conch shell for its head. For eons, giant turtles—the 120-pound green sea turtles and the 200- to 300-pound loggerheads—have laboriously crawled up on the beaches and made the area around Melbourne their favorite nesting site. Above the high-tide line, which they find without road signs, the females scoop out the sand and deposit one hundred or so eggs before returning to the sea. Covered by the warm sand, the tiny babies begin hatching seven to eight weeks later. Somehow, the babies know which way to go and begin making their erratic, perilous way down to the sea.

Nature has arranged for the crop of baby turtles to be thinned by seabirds, fish, and raccoons. Nature did not arrange for or count on vandals digging up the nests for fun. Nor did she count on so many people using the beach or contractors developing so much of the sandy area with bulldozers and pilings. So there may not be enough sea turtles left to support a genetically diverse population.

The brown pelican is losing its nesting grounds along popular waterways, too; and what does the manatee know from boats? One-third of the world's entire population of these slow-moving, walruslike creatures lived in Brevard County waters. Some believe the manatee accounts for the origin of the mermaid myths, which may say more about the eyesight or hormone level of ancient sailors than these lumbering inhabitants of the sea.

The manatee's feeding habits—one hundred to two hundred pounds of waterweeds a day—help keep the waterways clear of choking plants like the water hyacinth. This, however, helps keep the waterways full of motorboats, whose movements rarely take into account the existence of the 1,500-pound manatees. They sleep on the bottom and periodically float to the surface for air, which often brings them into painful contact with whirling propellers, which eliminate one more natural weed-eater. It is just so hard to remember all the things that think they own the land and the sea.

Boats are big in Florida, bigger even than in Mamaroneck. They're everywhere—in the driveway, in the yard, behind the car, or the camper, or the house. They're even in the water. The Intracoastal Waterway, that dotted line of a boat road that runs down the East Coast slightly inland to protect private craft from German submarines, is full of migrating boats every fall and spring. Some boat owners have made the journey so often, they simply must hire transit crews to avoid the boring repetition.

Some Florida communities even have Christmas parades with boats. Over many years developers have built feeder waterways in some Florida areas, honeycombing neighborhoods with little boat alleys, which happen to increase the value of property and can bring floating trash and water snakes right up to your back door.

U.S. 1 crosses numerous liquid alleyways like this, sometimes with a low drawbridge so that both boat and vehicular traffic can hold up each other. As it creeps down the coast through the stoplights of the communities with names familiar to baseball spring-training fans, the scent of salt and the sight of change is in the air. Near Sebastian (5,042 KINDLY PEOPLE AND 6 OLD GROUCHES), Elmer's trading Post now stands next to Yun's Produce, and one self-explanatory sign offers help through a phone number: CALL 636-BAIL.

Some communities have fascinating and revealing histories. Take Stuart, for instance, where pineapple growers objected to losing any land to something as ridiculous as a Florida Flagler resort, so the old man kept going. Or take Gifford. It was named for F. Charles Gifford, who is credited with having chosen the site for Vero Beach. Mr. Gifford is also credited with having delayed the extension of Mr. Flagler's railway by demanding an exorbitant price for a slice of his land. In appreciation, the railway founded a town exclusively for blacks, who were called Negroes in those days and were expected to live off by themselves. The company named that community for their neighbor, Mr. Gifford.

In the old days, more vehicles with out-of-state license plates seemed to get speeding tickets along Florida's Route 1. Not to imply there was any attempt to milk money from vacationers with only a limited amount of time to waste, but they did seem to earn more than their share of citations from the alert municipal motorcycle centurions whose salary more than likely came out of the Traffic Fine Fund.

With all of the traffic lights and the convenience stores and malls, there's not much chance to speed on U.S. 1 down there today. And this may be the answer to speed enforcement; build more malls, install more lights, and have everyone get on the road about 4:00 P.M. It's like New Jersey with sun and without the refineries.

While stalled in traffic, drivers can be entertained by the nearest bumper, now that some state supreme courts have ruled that stickers are a form of protected free speech: A WOMAN'S PLACE IS IN THE MALL, ENGLISH FOR FLORIDA, WORK IS FOR PEOPLE WHO DON'T FISH. Opportunely, at rush hour come the radio ads for cellular phones, which fight that dreaded affliction, being out of touch. With a phone in the car, drivers can put to use that formerly wasted downtime in traffic. Additionally, car telephones help avoid messages piling up back on your desk. Of course, we could do like the Japanese: Put a pay phone every few meters everywhere and have everyone take the train.

Even along the coast, Florida's U.S. 1 has its share of peeling paint and boarded-up buildings. Fort Pierce was an old army garrison during the Seminole disagreements when Zachary Taylor made such a military name for himself that he, too, could become an ex-general and president at the same time. The screen for the old Fort Pierce Drive-in is still standing, but the grounds now are covered with newly fallen logs about to become lumber. Just south of town, in Port St. Lucie, is where the New York Mets travel each spring (by air) for baseball training.

Jupiter before it was called Jupiter is where Jonathan Dickinson, a Quaker, and some of his party were swept ashore in 1696 after a storm collected their ship. They were captured by Indians, then freed, and allowed to wander some vague Spanish trails north a couple of hundred miles to St. Augustine. Jupiter is the sole remaining community on what was called the old Celestial Railroad, a line of tracks also serving the towns of Neptune, Mars, Juno, and Venus. That railroad left, too, after Flagler's through line went through and carried considerable commerce away to Miami. But Burt Reynolds, noted Georgian and adopted Floridian, keeps a house here, and, in commemoration, Jupiter has a little Burt Reynolds Park.

Some city must be drawing near because the tree lawns along Route 1 are becoming littered, the motels are small and not famous, and the factories, graphics plants, and parking lots have taken to surrounding themselves with tall chain-link fences garnished with barbed wire.

Serving South Florida Since 1988

IF MR. FLAGLER COULD see Palm Beach now, he'd turn in his grave—to endorse all the rent checks. And if none of the posh city's potentates travel by train anymore, they do park their private planes out at the airport, which should be far enough away that *they* need not hear their neighbors' jet engines. I mean really, could you imagine anything so vulgar?

Palm Beach owes its good fortune to an unfortunate shipwreck in 1878, which spilled coconuts into the surf. They bobbed onto the beach. Some took root *et voilà*. With a few billion dollars, you have a small city of perhaps ten thousand people. That city has a most impressive collection of yellow ascots, houses with names, old trees, women who use middle names, expensive stores, scenic views, and the kinds of swept driveways, spotless white cars, precise hedges, and trim gardens that grow only when some adult with only a first name is paid handsomely to do such work. Now, this is old money that they are spending. Well, they're not actually spending the old part; that gets divided at divorce time. They live off the new interest and dividends.

To be sure, there are a few newcomers. But you can spot that crowd. They're the ones who wear their hair over their ears and a bejeweled blonde on each arm. Take Donald Trump. Please. He bought Marjorie Merriweather Post's ornate 118-room mansion, Mar-a-Lago. But next thing you know, he'll want to subdivide the grounds and sell mini-estates. You watch.

Mr. Flagler's modest fifty-five room mansion, Whitehall, is a museum now, complete with the secret staircase he used to escape all the guests one of his wives kept inviting over.

In most of Florida, "old money" means someone who got rich in the late eighties—the *1980s*. However, most real Palm Beach homeowners were millionaires in utero. They were raised by servants to do things like sign their names and ride horses. Guess where Prince Charles goes to play polo after a hard day of cutting ribbons and greeting guests. Oh, and croquet is still big there, too. The county has eighteen croquet lawns. In fact, the U.S. Croquet Association's National Open Club Team Championships were just played in Palm Beach. And the Challenge Cup, which was conceived to settle that burning question of who plays better croquet, the British or the Yanks? (Wasn't this settled some time ago back up around Yorktown?) The brunch to watch those croquet finals costs only twenty-five dollars.

Palm Beach is the kind of place where alleys are called Vias and where a bumper sticker—SHOP TIL U DROP—wouldn't even draw a smile. First of all, bumper stickers are so, well, common. And sticky. And second, what's tiring about shopping the length of Worth Avenue when you pay people to follow along with the car and collect the boxes? Or do the grocery shopping at *the* supermarket in town, which has a nice selection of wine for under $299 a bottle.

Winter is the busiest, I mean, the ultimate busiest, time of year in Palm Beach. There are just so many things to be seen to be doing, charities to help, arms to pat. Although no one actually does anything quite so brazen as claim it, it is said that the Venetian Ballroom at Mr. Flagler's grand old frescoed, tapestried Breakers has seen more money change hands for charity than any other such room in the country. That is the kind of claim that can be made comfortably with the knowledge that no other place is counting. But the fact is that Palm Beachers do have a good time doing good.

Palm Beach sits on the northern end of a tony island, which locals would probably prefer to call Anthony. It is connected to the mainland by three bridges over the Intracoastal and Lake Worth (Bill Worth was a general in the Seminole War). The mainland is West Palm Beach, which doesn't have a sidewalk drinking fountain for dogs next to the F.A.O. Schwarz store because it doesn't deserve an F.A.O. Schwarz store. West Palm Beach was originally intended as an affordable community for the help—the gardeners, maids, and drivers, who were not qualified to live in P.B. proper.

So West Palm Beach is where things like laundromats are. Laundromats, you see, are not allowed in Palm Beach. This could be because Palm Beach residents throw their clothes away after one use or because Palm Beach residents do not get dirty or sweat. (Of course, they might have their own washing machine, but that would be in one of the back rooms, so who would know?) West Palm Beach is also home to such brightly lit lots as sell used cars, which are called "preowned" thereabouts.

Even in its shiny early days, Route 1 was not good enough for Palm Beach. It had to settle for passing through West Palm Beach. Highway 1 becomes more a road through southern Florida's heavily developed gold coast. Towns, even cities, come and go seamlessly. Just as it's hard to find Route 1 in northern cities, down there it's hard to find the cities on 1. Where does one end and the next begin? Are, for instance, the Caribbean Motel, the Breezeway Motel, or the Sun-Ray Motel in the same town as the Cadet Motel, the Wishing Well Motel, Shane's Motel, or Johnny Mango's Produce Store. And if so, who cares? The headquarters for the *National Enquirer* is nearby on 1, as is Coastal Used Cars (SE HABLA ESPAÑOL) and Morey's Lounge (TOPLESS GO-GO GIRLS!).

It is not, however, hard to tell when one enters Boca Raton. There's a Flagler (him again!) National Bank, a name conjuring up images of power ties and wealthy mustaches in the older Florida psyche. For the newcomers, Flagler seems like one of these defeated Confederate generals they're always naming things after down there. And then there's Elena's Bikinis, which is what you call a swimsuit store in a place like that. Boca has been around since the mid-twenties. With an altitude of seventeen feet above sea level, it is the kind of place where officials nodded and agreed unanimously with the lawyers present that it certainly made sense to cut through and dredge out a sizable chunk of land to let the ocean into Lake Boca Raton so that some Boca folks could get their yachts right up by the club.

Boca ain't cheap, but it ain't Palm Beach, either. It attracts an upscale, well-dressed crowd of Northerners, some very northern. There is, on any given winter's day, one-twenty-sixth of Canada's entire population in just that one state, not counting the Canucks in Arizona and California and Hawaii and anywhere else it's warm. CANADIAN SPOKEN HERE, say the shop signs. The Canadians are even getting a little uppity now and then, at one point trying to organize and force local Florida stores to accept Canadian dollars at par with U.S. dollars. At times this would have cost the store owners twenty-five cents or more on the dollar. And anyway, Canadian money is funny-colored.

But unless they're waving a brown two-dollar bill with a woman's picture on it, it's not that easy to spot the English-speaking Canadians, if they're not talking. They do say "eh?" a lot, as in "Going to the beach, eh?" They are polite. They are terrible drivers, by American standards; this means they don't drive like ambulance operators. They have their own weekly newspaper.

Les Quebecois live apart from the English-speakers, as they do back home, too. They have their *own* parties, their own weekly newspaper, and their own Florida beaches like Hollywood, where English is a foreign language. About the only time that these two North American solitudes get together is on winter Saturday nights at the Penalty Box. That's a Fort Lauderdale bar on U.S. 1 where the satellite TV screen is as big as Manitoba and everyone can cheer for their favorite 'ockey teams, which have yet to make it on to American network television, to the eternal puzzlement of Canadians.

According to police, the annual invasion of Canadians even includes those somewhat to the left of the law. Canadian criminals have been known to pull a heist in, say, Montreal in November, and then spend the rest of the winter in Florida to avoid the legal heat back home. Southern Florida has been an isolated wild American frontier for a very long time. First, it was due to pirates, the Spanish, the Indians (or, if you were an Indian, the army), and, ultimately, real estate salespeople, who keep growing back like weeds, despite the early boom-and-bust cycles of Florida's famed and sometimes fraudulent land sales.

For one thing, it was very hard to get around in Florida, except by water. The U.S. government didn't even try to police the territory at first. It just built a series of refuge houses every twenty-five miles along the coast where each season's refugees and shipwreck survivors could hole up for a while. During the Seminole Wars, the army carved the Capron Trail out of the jungle between the tiny settlement of Miami and the Fort called Lauderdale (name sound familiar?) after a Tennessee captain (first name William). The Indians believed that the unusually deep New River there (it wasn't old) was created over one night by spirits; archaeologists have since decided that may well be true, the result of an earthquake splitting open rock covering a vast underground river. The waterway was ultimately great for fishing, and the alligators liked it, too. The army was there trying to cut off Indian supplies from Cuba (today, the National Guard is trying to cut off drugs from Colombia). There were several massacres, including one led by a chief named Sam Jones-Be-Damned, who also ordered parts of the ears cut off Crop-Ear Charlie (that was his name after, not before) as punishment for tipping off some settlers about an upcoming raid.

But after the Indian fighting, that army path was quickly overgrown. And for many years until nearly 1900, anyone wanting to travel from Palm Beach to Miami by land walked the sixty-six miles of beaches. (On some modern days, that route might still be quicker.) For safety in those days, such travelers walked along the beach with the barefoot postman, paying five dollars for that privilege, because the postman knew where he had hidden all the little skiffs necessary to cross the myriad streams and eddies along the way. Over the years, some stages struggled through the sand, but it wasn't until 1903 that a rock road was built onshore. And today's Route 1 follows those old paths, past jungle nursery growth carefully planted to give the mall areas that tropical feel after the original jungle growth was torn out to provide that familiar bulldozer feel to the landscape.

Much of this growth has come only in recent years as the spaces between previous developments were filled in by a modern generation called condos. One of the early architects was Addison Mizner, who bought and designed several thousand acres of farmland with the dream of a Boca Raton resort. The big real estate bust of the 1920s shattered those hopes (and a pair of massive hurricanes didn't help, either). As late as 1950, Boca's population was 992. Today, it's passing 50,000 and two of those residents are the Vecseys.

George and Marianne live part of the year up U.S. 1 in New York State and part of the year in their beach-front condominium apartment just off the same highway in Boca Raton. When they're not in Florida, they rent it out, not a bad investment at all in recent times. Their official purchase of the apartment was set for an October morning at a local realtor's office.

With her husband away on business, Marianne drove in on the moonlit evening before with her car loaded with clothing, valuables, and documents. The first motel room she examined was dirty, so she checked into another. But, arriving in that second-floor room, she noticed that the security chain was missing. New Yorkers instinctively consider such things a necessity. So, carrying her camera, a radio, and her purse with all her money, travelers checks, and the certified check for the apartment, she went down to the desk clerk to complain. There was a long line of check-ins. She went back to her room to wait a few minutes.

As she approached her door, she heard running, which seemed strange at that time of night. She usually worked out in daylight. As she opened her door, she turned, and a black man pushed her down into the room. "It was just like in the movies," she recalls thinking.

Without thinking, she fought back. She kicked him in a delicate place, scratched, screamed. He stepped back, startled. She screamed some more. He looked down the hall, then grabbed for her purse. She knew what was in it; he could only guess. They both yanked on it; he wrenched it hard and sent a sharp pain up her left arm.

Panicked, he let go of the purse, grabbed the camera, and fled. "That's all he got, the camera."

Within three minutes, the police were there, asking questions, getting a description. Marianne's adrenaline was still flowing. Her daughter, Carianna, had been mugged in Brooklyn just two months before and, also not seeing a weapon, had reacted the same way. "Scream-ing really works," Marianne kept telling the officers, who prefer other tactics.

Late that evening, they drove her to the hospital emergency room, where Boca Raton's social set does not usually gather at 11:00 P.M. With her casual driving clothes disheveled, her hair windblown from the day's drive, and her hand seriously swelling, Marianne drew curious looks from the staff, who asked, "Have a fight with your boyfriend?"

The next morning, Marianne attended the closing, got the keys, and then went to a hospital, where a pin was inserted in her broken fingers. As the investigation continued, Marianne learned from the police that such muggings were no longer rare. The criminals, usually drug addicts, seem to come out of Miami up Route 1 or the Interstate to a row of motels, where their victim candidates are less likely to be familiar with the people and the dangers. There had been many such attacks, some quite violent. Typically, the muggers jumped in their cars and escaped back down the expressway to the anonymity of the city. The police seemed amazed that Marianne had resisted, and dubious of its value.

Some weeks later back in New York, Marianne got a call from the Boca Raton detective. He had faxed an artist's sketch of the crmiinal based on her description. An artist herself, Marianne sent her own version back. And on a trip to Florida soon after, she was summoned to the police station to view mug shots.

"That's him!" she shouted. "That's him!"

"Relax," said the detective. "We've got him. And his partner."

The pair had been caught beating an elderly couple unconscious in a parking lot. They were drug addicts. They did live in Miami, though not anymore. Eventually, the men were charged with sixteen counts of assault. Mrs. Vecsey's attacker plea-bargained on the charges, getting the potential total sentence lowered from four hundred to ninety-nine years.

He's one of Route 1's growing number of prisoners now; he'll be a guest of the state at least until the year

2020. That makes Marianne feel less jittery, although the mugger said he had no particular memory of her incident. Marianne's fingers are better, though one is slightly crooked. Her hand aches often, especially when it's going to rain. "But I'm not blind and I'm not dead," she said, sounding like many urban American victims who begin appreciating things that other generations took for granted. "I know much more now about what goes on in this modern world of ours. I'm more cautious about people and places. I don't assume the best of strangers anymore. I'm wiser."

Turning inward, guarding the gates of involvement, looking out for Number One, fortifying personal defenses against a threatening outside is a social movement that many criminologists have noted among Americans in recent times. "It's simply safer," said Frank Hartmann. He is executive director of the criminal-justice program at the Kennedy School of Government back up Route 1 in Cambridge, Massachusetts, which has its own unsafe times; Mr. Hartmann's car was stolen recently. This kind of event, repeated millions of times and officially reported to police barely half the time, has created a critical mass of fear and suspicion. It has created an invisible fraying of the social fabric, which makes each individual feel apart, accountable only to themselves. This is, after all, a free country, so everyone should be free to be afraid, too. This alienation and isolation is strongest in the larger metropolitan areas, but not confined to them. No longer, for instance, do many big-city residents watch their neighbor's house, if, in fact, they know their neighbor. No longer do they dare to correct noisy teenagers or personally question some individual unfamiliar to their street. Too dangerous.

The main result is that the role of police has changed. Originally, in Britain, the constable of police (COP) was in a stationary office where he accepted complaints from citizens who produced the accused, as well. Peacekeeping was a function that required every citizen's participation, like voting, which has been going down, too. Today,

American police have become merely another urban utility. You pay for it. You receive access to their services via 911. You're not involved. Lawbreakers are someone else's problem.

This system seems clear. It seems crisp and safe and compartmentalized. Problem with the pipes? Call the plumber. Problem with the power? Call the electric company. Problem with crime? Vagrants? Noise? Litter? Call the cops. But this has proven woefully ineffective, not so much a system as a failure of one. If every other person was a police officer, this might work. But no amount of cops and tough sentencing laws can make up for urban poverty, decay, discrimination, drugs, and a generation of mindless, often parentless, and discipline-free child care—as is witnessed by rising violent crime rates virtually everywhere. Washington, D.C., where hookers and drug dealers ply their trade within site of the president's house, has also become the nation's capital of homicide. Farther up Route 1 in New York City, the average day sees about six murders, over two thousand every year, despite the presence of a uniformed force of police equivalent to two full army divisions.

Nationally, there are a half million law-enforcement officers, and about twice as many prisoners. About 150 officers are killed every year. Police protection costs Americans $28 billion every twelve months, about $114 per person, or barely 8 percent for domestic defense of what each American pays for military defense.

One of these local defenders is Glenn Maura, a twenty-four-year-old Miami officer who patrols that city's deeply troubled, black Overtown area, scene of some deadly riots not long ago right alongside U.S. 1. Five times in the last ten years, Miami neighborhoods have seen an outbreak of such mob violence. That is not counting the uncounted number of street-corner drive-by drug shootings, which criminologists have taken to calling "quiet riots."

Such outbreaks are attributed, familiarly, to many factors, among them the area's persistent poverty, the scourge of drug addiction, and the accompanying crime cycle that reaches out to touch nonresidents such as Marianne Vecsey. There is, too, a history of racial discrimination and frictions heightened by a newly emerging immigration pattern in American cities. While numerous older large cities were losing population in the 1980s, the Miami–Fort Lauderdale metropolitan area was growing by 21 percent to 3.2 million. Immigration from such places as the Dominican Republic, Cuba, India, and the Philippines produced this growth, as it did back up U.S. 1 in Elizabeth and Jersey City, New Jersey. The Hispanization of Miami is such that even the old Miami *Herald,* another chain-owned newspaper that sits alongside Route 1, offers Spanish-language editions.

As part of his job earning about fourteen dollars an hour, the six-foot-three Officer Maura wears a bulletproof vest and must take Spanish lessons. Working the 3:00 P.M. to 1:00 A.M. shift, he also gets informal lessons in cynicism. "You risk your life out here for what?" he says. "You grab some bad guy. The judge slaps him on the wrist. He does no time. In fact, he's back out before I finish my paperwork on the arrest."

The son of a customs officer and a nurse, Glenn was attracted to police work by the idea of public service, the satisfaction of helping people even if, full of fear and sometimes pain, they don't always remember to say thanks. "But all we do out here is respond to 911 calls," he says as he writes up a report on a woman's stolen license plate. There is no hope of ever recovering the plate. She will have to take a day off from work and go down to the motor-vehicle department to seek a new one. But Glenn's report is required by the state bureaucracy to certify that the plate is really gone. It consumes thirty minutes of his shift. "There's no time to do any real police work."

Confronted by their ineffectual crime-fighting and crime-prevention programs, a growing number of communities, even New York City, are experimenting with the so-called community-policing concept. In essence, it is an attempt to return to the effectiveness of the old days by giving officers an assigned beat and a sense of own-

ership of an assigned area, thus freeing them from slavishly responding to 911 calls, many of which are not really emergencies. Thus, the officers become more familiar neighborhood problem-solvers; instead of simply responding to mugging call after mugging call at one dark corner, for instance, they invest time in solving that problem by doggedly fighting the city bureaucracy to have that corner's streetlight repaired or installed. The problems are deep, of course, and so is the cynicism of some law-enforcement professionals who have learned to stand aloof as self-protection. By some statistical standards, Miami's crime rate is among the country's worst, with much of the flavor of drugs and violence that are more vivid in real-life than their pastel presentation on television. But formally implementing community policing is not a Miami priority at the moment in Dade County, which is named for Major Francis L. Dade, an early homicide victim.

While Glenn's daughter Temica sleeps peacefully in their apartment ("Someday soon I've got to get a house"), Glenn spends his time cruising the ghetto's dark streets, stopping suspicious cars (weaving, missing lights, expired plate tags), breaking up a couple of fights (usually, beer is involved), and sitting to watch a handful of known drug dealers stand on a crumbling corner, awkwardly fidgeting beneath his stare. It's all part of the constant pushing match, each side trying to intimidate the other. Several times, his random patrol paths cross those of colleagues, and they stop their cars, driver's side to driver's side, to chat and trade observations. At other times, they exchange coded messages via computer screens right next to their steering wheels. On the side of the computer console is an orange emergency button. Pushed, it will silently radio his location and summon help.

At one stop, he grabs a chicken sandwich in case the evening gets busy. He eats so lightly ("Maybe it's nerves") that he can't put on any weight, even with the weight lifting. There's a sex offense to handle at a local hospital. With the patrol area's lowest seniority, Glenn draws that distasteful duty, interviewing a frightened youngster about who touched him where and when.

At one convenience store, he does a little inadvertent public-relations work. A young black boy engages him in conversation about Glenn's nine-millimeter pistol, and then asks if the officer knows of any jobs where he could earn honest money. Glenn is touched; he promises to check around and let the kid know next time they meet. He pulls out some paper and makes a note. His belt is like a walking closet—pistol, ammunition, can of Mace, loop for a club, flashlight, and some tough plastic thongs. Throwaways, they are a safer kind of handcuff for the age of AIDS when no one wants any unnecessary blood splashing around anywhere. He also has a special protective mouthpiece should any mouth-to-mouth resuscitation be necessary.

Just before midnight, Glenn's beeper goes off; he checks the number and smiles. It's his wife, Cassie, making her regular security check before she goes to bed. She likes to be able to reach him, though she rarely wants anything other than to hear that he is okay, so far. He phones from the next convenience store.

Around midnight on U.S. 1, Glenn encounters a nice-looking car, which is suspicious in the impoverished Overtown area. No one enters the black area on purpose unless they live there, are assigned to patrol there, or are up to nothing good, like buying drugs. Glenn pulls up alongside. Four very fresh, very white teenage faces look over and smile.

"Man, are those girls lost," says the officer.

"Can I help you ladies?"

"Yeah," says the driver, who's trying to return to the suburbs from a rock concert downtown. "We're lost. Where's the Interstate?"

It turns out to be not a busy day. Glenn didn't even draw his gun once, though he did keep his hand on it a couple of times. "It's the time you're not prepared that you get in trouble," he says, sounding like a veteran after only fourteen months on the force. He stays safe by a simple philosophy. "I'm not the Rambo type. But don't tell any of the bad guys."

By 1:30 A.M., Glenn will be in the police gym working off some of the stress he is not to show on the job. He'll go home, watch some early-morning TV, and then go to bed until early afternoon. By 3:00 P.M., he'll be back out on U.S. 1 again in the same car looking for the same things, answering the same 911 calls, completing the same reports.

Someday, for a change, he'd like to get into narcotics enforcement, going undercover, making the kind of street-level buys and busts that never make the papers, unless the undercover cop is uncovered, injured, or worse. But for now, there's a coffee break with two other officers at a late-night sandwich shop over on the main highway. The night clerk is frightened, standing alone in his lighted shop at this hour. He believes that police, even those just strolling by or sipping coffee, are a powerful deterrent to trouble. The man is always so glad to see uniformed police entering his empty establishment at unpredictable times that he offers eager smiles and free refills to each patrolman.

"Hi!" says the clerk, who does not know the officers' names. "How you doing tonight?"

"Fine," reply the officers, who don't know his name, either, but still sharply check the place out with their eyes, lest their arrival has interrupted something unpleasant. One of them turns on his portable radio so they can hear all calls for trouble. Ten minutes later, they walk back out the door onto the urban sidewalk of U.S. 1.

When they leave, the clerk leaves. The shop lights go out. And another piece of the old highway goes to sleep for another night.

The Bottom

IN FLORIDA'S FIRST CENSUS almost 175 years ago, 317 people were found living in Florida south of St. Augustine—not counting Indians, of course, who didn't count in those days. The population included some characters like Old Cuba, a refugee from there (even in those days) who was short, liked machetes, and got eaten by sharks, and Bill McCoy, a teetotaling rumrunner whose high standards for hooch gave rise to the expression *the real McCoy*.

There may be fewer genuine characters these days, but there are certainly many more people. At some point, most of them seem to cross Route 1, be they en route to Pure Platinum (FEMALE BOXING AND OIL WRESTLING), the dog-racing tracks, or jai alai arenas, to Park's Clams, the huge Humana Hospital, the Baseball Card Mania shop, or Braman's Used Car-Cadillac-BMW-Rolls-Royce showrooms (Norm Braman is a Floridian who went back up U.S. 1 to buy the Philadelphia Eagles). The road becomes part of a high-speed-auto racecourse one day every year. It winds past the brightly lit cruise-ship docks, now the world's busiest thanks to Henry Flagler's dredging. The pavement winds alongside Biscayne Bay, the broad waterway that divides Miami from Miami Beach and is home in winter to upwards of 40,000 pleasure boats. U.S. 1 then straddles Miami's palm-laden downtown median strip, navigates a couple of corners to pass the luxurious major hotels where the lobbies are empty all night save for the clerk ringing up everyone's bill to speed up the early-morning dash to the airport. Miami's nearly 650 motels share a tourism market of nearly 7 million visitors every year.

Just south of downtown, the highway brushes past an array of graceful old Miami homes now become bank offices. It also passes Florida International University and Miami University, which does not pit its sports teams against the Ivy League schools, such as Brown and Princeton, that line the same road back up north.

Next to Miami is Coral Gables, which is home to about 45,000 people, and is another one of the state's relics from the optimistic age of boom before bust. Coral Gables was the idea of George Merrick, another son of a northern minister, whose imagination flowered in the seventy-eight-degree days of south Florida. Merrick the Younger had the idea of creating a sub-urb, a community right next to a major downtown where people could live luxuriously but still conveniently. Beginning with orange, grapefruit, and avocado groves, he assembled a land empire, which he subdivided into planned residential, commercial, and recreational zones built in the Mediterranean style. By 1921, he was ready to start selling. There followed several dizzy years of hot sales, skyrocketing prices, speculation, and big profits. The real estate bust of 1926 presaged the upcoming doom of 1929. The hurricane of September 18, 1926, with winds of 110 miles an hour, made the situation worse. Merrick went bankrupt (so we don't need to call him Mr.). Pieces of his municipal dream broke off. Merrick and his wife moved down the road to Matecumbe Key and ran a fishing camp. Since those were the days before Watergate, President Roosevelt could still name a bankrupt real estate broker as Miami's postmaster in 1940.

Merrick died two years later. (But Coral Gables's Miami University has had a pretty good football team in recent years.) And with the growth of the private automobile, the American concept of the suburb was safely planted in the nation's psyche to blossom and flourish another day. The 1990 census showed that, for the first time, a majority of Americans live within about forty metropolitan areas of more than a million residents. These are the ever-sprawling areas away from the increasingly crowded and undesirable cities. The jobs are growing in suburbia because that is where the skilled labor is, not to mention the cheaper land. This all sets up a complicated new dilemma for the next decade as the jobs that once drew the affluent into the old city now follow the affluent out of the cities instead, leaving the unskilled poor to fester in unemployment or become reverse commuters to areas weak in public transportation.

The massive movement to suburbs is happening in other lands, such as Japan; the difference is that there a modern mass-transportation system is in place. The idea of mass transit in many American areas is to declare one expressway lane open only to cars carrying more than one person.

The eminent historian Daniel Boorstin likens the American suburb to the old wagon train as a transient, truly American community where residents leaned on one another as they moved about the country and up the scale of living. A small town was where people stayed, where roots grew and values formed. A suburb was a temporary home where people lived until they moved to a slightly better-grade suburb somewhere else. And everyone assumed that the children, while they might move to a city apartment in the initial years of their adult life, would eventually end up in a similar though different and equally comfortable suburb sometime later. They were like McDonald's, convenient, largely interchangeable, and familiar, even if you'd never been there before.

From aging Coral Gables, U.S. 1 ducks by the Tamiami Gun Shop, Andy's Camera House, Tile City, Carpet City, Stereo City, and Circuit City water beds, breezes by a banquet of such food emporiums as Pig City Barbecue, Wings 'n' Things, El Taco Rico, and parallels the Miami area's elevated Metrorail transit system. On the way, it passes West 216th Street, not the most evocative address for a place called Monkey Jungle. That is a twenty-acre wild reserve for monkeys founded by Joseph Dumond some sixty years ago. It is an intriguing concept: Humans stroll through in caged walkways while all kinds of monkeys, running free, watch from the trees. (But it's

the humans who pay for this privilege. So, you figure which is the smarter species.)

Then, U.S. 1 enters Homestead, home of the air force base that was such a vital link in ferrying planes to Europe and Africa in World War II. Karate College is there, too, and Orchid Jungle, another of Florida's unlikely success stories. (But wait till we get to the old boat in Key Largo.)

Homestead got its name from its homesteaders, who could gain title to 160 acres of the rich land by living there and farming it. The city has been the commercial hub of a productive agricultural area, along with its neighboring community with the really imaginative name of Florida City. (Original settlers wanted to call the town Detroit, but they couldn't produce enough murders to live up to that name and the post office anticipated considerable confusion amongst its literate letter carriers, so the name was changed.)

One might say the Homestead area is suitable for agriculture. It has more than 350 frost-free days every year, which is a pretty good start. And it has seventy-five to one hundred days of over ninety-degree weather, which is also pretty good, if you're a plant. Unlike Johnson's Florists back in Philadelphia, Orchid Jungle grows its own flowers. It began in 1888, in Kentucky, as the business of the Fennell family breeding new hybrid orchids and clones of the lovely tropical flowers. In 1922, the family moved to Florida with the land boom to grow the plants in their native setting, which began attracting a few local residents, who brought their visiting relatives, who told their friends back home, who came on their next trip, and so on.

The Florida Keys are a string of tropical islands running for about 125 miles from the eastern shore of the Florida peninsula in a gentle curve around to the southwest. They were once connected only by the azure blue sea. Then in the early 1900s, along came Mr. Flagler's tracks, which stitched their coral mounds together with lengthy overpasses above the sea all the way to Key West. There, the railcars were shunted onto boats for the ninety-mile journey to Cuba in the days when that watery gap was not augmented by an ideological one. Then on September 2, 1935, along came another great hurricane, which removed forty miles of overpasses. The railroad opted not to rebuild, transferring its Cuban ferry service to Fort Lauderdale. And the state took over the watery Keys right-of-way and built the first of two main highways. The second was a wider, sturdier version that replaced thirty-seven narrow bridges in 1982 at a cost of $185 million.

The road's mileposts are vital markers since, in the Keys, directions and addresses are given not by street number (there's only one main street, anyway) but by mile marker, which begin outside of Florida City at 126 and run down to 0 at the bottom of the United States. Parts of the crumbling old vehicle road still stand by themselves today in the warm waters with no beginning and no end. The old sections of pavement are now mere pelican roosts equipped with railings for fisherpeople to lean on while their scaly prey below examine the hooked offerings.

I don't want to suggest that, even with the U.S. 1 connection, the Keys remain somehow psychologically apart from the rest of the United States. But the highway has a few LAST CHANCE signs, as if motorists were about to drive into the Unknown Zone. The gun shop at Key Largo has a small guided missile out front, presumably for display purposes only. A large billboard now ominously announces to travelers, DOMESTIC VIOLENCE IS A CRIME IN MONROE COUNTY. As if it wasn't always. And the post office at Tavernier, a favorite hiding place of the late pirate Jean Lafitte, today has two large blue letter-collection boxes outside: One is marked KEYS; the other is labeled WORLD.

It is a fairly desolate stretch of highway down to the Keys from Homestead: two lanes, dead trees on each side, the muggy scent of sea air hanging about. Several of the tall pylons carrying electricity out to the Keys now double as high-perched homes for large seabirds, who ferry twigs, grasses, and even sizable branches up to the metal crossbeams and, there, do their lookout duty and raise their

families in open-air penthouses. The state, aware of the highway fatality rate on the often-crowded and slow-moving throughway and of modern drivers' strongly felt need to rush, has erected its own series of Burma-Shave signs with one word on each board: PATIENCE PAYS ONLY THREE MINUTES TO PASSING ZONE.

A small-town flavor survives in the Keys. A photocopied sign on several posts pleads, PLEASE HELP FIND MY BEST FRIEND. It offers a three-hundred-dollar reward for finding the three-year-old parrotlike bird named Elex. But a lot of visitors do come (or else the locals are really crazy for omelets). WELCOME TO FABULOUS KEY LARGO, announces the sign at Kuntry Kitchen (get it?), WITH PLEASURE OVER 1,514,016 EGGS SERVED.

Miriam Otera is one visitor who came and stayed. Another native of Toronto, she married an American twenty-five years ago and they set off in their twenty-five-foot boat to sail into the sunset. Actually, they went to the East, all the way around the world. It took them seven years. But they were in no hurry. And when they dropped anchor for the most recent last time, they were in Key Largo.

That was eighteen years ago, many cycles of tourist seasons. The Oteras still live on the same boat. Miriam works down U.S. 1 at the Rain Barrel, a local crafts store built around a lovely outdoor garden with a dense collection of shady trees and vines. There is considerable piecing together of an economic existence in the Keys, a little of this and a little of that put together with some of this, and a body can get along comfortably. No need to buy parkas, just air for the diving tanks, if you're into that.

Many people are into owning a piece of this and that. As one result, Florida has more than its fair share of developments called Sandy Cove, and they're not named after a baseball pitcher. As another, the slower old ways are melting away. The speed limit goes up. More people rush around faster, getting less done but more quickly, and giving some the distinct feeling that Waylon Jennings is right when he sings about society climbing ladders to a hole in the ground. "You wouldn't believe how peaceable and quiet it was here when we arrived," said Miriam, who was doing without many customers that morning. Local artists, of whom there are many, bring in their pottery or paintings or whatever, and the Rain Barrel tries to sell them. Does a good job, too. Not as brisk a business as the all-American, free speech, walking billboard T-shirt industry over at the Largo Cargo Store. (By the way, T-shirts of this book are available in the lobby, along with original sound-track recordings.) And there's been so much building everywhere in the Keys. Hardly any open spots left, the locals say, looking out on the vast expanse of ocean that surrounds them on all sides, striking visitors as magnificently open. The Keys's sole link with the world is those two lanes of cement of 1—plus the TV, of course.

Quite an invisible split has developed between the old-timers and the newcomers. (Sounds like Maine, doesn't it? Except Maine's idea of old is grayer, less sunshine. And the water is maybe forty degrees cooler.) The Upper Keys's population of around 20,000 doubles come winter, which creates a feeling of crowds and confinement if your livelihood has nothing to do with women wearing Day-Glo halter tops and men in plaid shorts, with shirts whose buttons must not be working, because the whole world is exposed to their hairy chests and bellies. Thank you very, very much. The winter heat also seems to cause the radio stations to play a lot of Beach Boys music. Happiness, goes one saying among nontourists, is seeing a Canadian going back up north on Route 1 with a New Yorker under each arm.

However, the visitors are great if your business is tourist-oriented like that of Jim Hendricks or Donna Casey. Even the Pizza Hut has begun accepting out-of-state checks. Donna runs the Florida Keys Welcome Center in Key Largo, the largest of the islands, which was a popular place even before it became a movie and a song. Donna, too, lived for years on a boat with her husband, John, a retired Central Intelligence Agency analyst. They had fled down 1 from the developing chaos and filth of Washington to Fort Lauderdale, which then got crowded

and developed, so they went a little farther south. There's not much farther south they can go and remain in the United States, so they bought a house, fortunately just before the values skyrocketed. But even owning the house, Donna had to go back to work to cover the cost of living where everything except some fish must be trucked one hundred miles or more. "Route One is our lifeline," she says.

There are real problems with all these people, though. The old-timers know what the Keys once were and they miss it. The newcomers from outside know what they're vacationing from. They don't miss it, and the Keys today look so pristine and pure. How could their second house or their garbage or their boat make any difference? This leads to today's definition of a conservationist: someone who built their second home last year.

There are a lot more boats now whose passengers often chuck their trash overboard (one state sign—DON'T TEACH YOUR TRASH TO SWIM—depicts a fish caught and drowned in a loop of a plastic six-pack carrier). There are more divers, who may bring things up from the bottom, although large parts of the area are undersea refuges. In fact, Key Largo's seventy-five-square-mile John Pennekamp Coral Reef State Park is the country's first underwater preserve. It is the result of Dr. Gilbert Voss's observations in the late 1950s that the area's entire conch population had been wiped out by souvenir hunters who turned the beautiful large shells into doorstops and useless shelf decorations.

But all the new building and cement makes for more runoff, too, which makes for trouble within the delicate symbiosis of the coral formations. "Coral reefs grow upwards," advises one state pamphlet. "When they are relatively close to the surface, the coral formations will make the water appear brown. Such areas should be avoided." That is a real good idea to keep keels in one piece. Naturalists call all this "human-impact activities." But running aground also destroys the reefs. And some boaters blithely toss their anchor overboard and rip out a whole section of coral. It's just rocks, right?

The reefs are up to seven thousand years old. They extend from up near Fort Pierce all the way down and around past Key West to the Dry Tortugas, which was the site of a federal prison where Dr. Samuel Mudd was sent for setting John Wilkes Booth's broken leg after the Lincoln assassination. (There never was any substantive proof that the Maryland doctor knew what his patient had done, but he did give the English language a lasting expression: *Your name is mud.*) The coral reefs can grow from one to sixteen feet in height in one thousand years. They provide shelter, food, and breeding grounds for countless plants and fish, which is good for them and also provides the attraction for tourists. While they look like rock, the reefs are actually a vast array of microscopic plants living within animal tissues. The plants draw nutrients from the animal wastes, while the animals gain from the energy produced by the plants through photosynthesis in the sunny waters.

The marshes, too, have been affected by all the development. Even if they're not drained, their water quality declines as the collections of mangrove trees die and give way to pines and condos, real estate that produces some real-time income.

There are also mounting problems within the 502,000 acres of Florida's sea grasses. These plants help oxygenate the water. They help stabilize the sea bottom with their roots. They provide sanctuary and food for all kinds of sea life, while providing a safe place for aquatic life to mature safely in the wild world undersea. And, by trapping fine floating particles, the grasses help keep the water clear, which allows the light in, which allows all kinds of life to continue. But the waters coming out of the freshwater rivers are carrying some interesting chemicals and effluent that don't help the grasses. And there's been a lot of absolutely necessary dredging and filling going on, which has wiped out large sections of growth. But, of course, it's just grass. And it's underwater, anyway, so who misses it? Some of the grasses grow above the water, so they are picked to make swell bouquets on someone's chipped Formica tabletops back home. UNLAWFUL TO PICK SEA OATS, reads the state sign. But maybe it wasn't big enough to see.

And still the people come, by the thousands. The Key Largo Chamber of Commerce even has a twenty-eight-minute color video to send to prospective hotel tenants and restaurant diners. Sure, sure, there are glass-bottom boats and colorful reefs and diving guides and charter boats to go after fish the size of a Honda. But one of Key Largo's big attractions now is a dumpy, rusting old boat that Jim Hendricks keeps tied next to his Holiday Inn. "The old tub needed a home," says Jim.

But this is not just any old tub. This is the mother of old tubs, the famous tub: the one and only *African Queen*. Yes, movie fans, *The African Queen*. This is the story of the boat of that story.

It is, to me, sadly symbolic of modern America that *The African Queen* is known as a movie instead of the C. S. Forester novel it originally was. But, then, I'm not writing a movie here, am I? And no one can dispute the fame of the rusting thirty-foot open-decked craft that carried Humphrey Bogart (and Katharine Hepburn) through vines, leaches, and waterfalls down perilous African waterways to Bogie's only Oscar.

The craft was built in Britain in 1912, named the *SL Livingstone* for the African explorer, and then shipped by steamer and railroad to work the waters of Lake Albert. For a few decades, the *Livingstone* anonymously chugged its way with supplies and pioneers and outlaws up and down adjacent rivers. In 1935, the Forester adventure novel was published, depicting the odyssey of the *Queen's* skipper Charlie Allnutt and his stern, prime, and puritanical passenger, Rose Sayer.

In 1951, according to detailed records collected by its current owner, the *Queen* was discovered by an assistant art director of the movie company. He had its original steam engine restored and the craft transported to the Belgian Congo (now Zaire) for its date with cinematic history. In the climax of the screenplay co-scripted by

James Agee and John Huston, the boat is sunk in an encounter with an evil German ship. But that was a mock-up.

The real *Queen* languished in Africa for nearly two decades before being bought for seven hundred dollars by an American film buff to begin a brief, less than glorious career as a tourist attraction on the American West Coast. In 1981, Jim Hendricks read of the attempted sale of the boat. He fell in love, sight unseen.

Hendricks had been on an odyssey of his own. Born up U.S. 1 in New Jersey, not far from the fabled Pallisades Park by Fort Lee, he had grown up in Kentucky and was living a hectic life as a Louisville lawyer when, at the age of forty-two, a heart attack gave him a premonition of mortality. He not only slowed down, he changed careers. He bought a local Holiday Inn franchise and then traded it for the one in Key Largo, where Bogart also had spent time making the movie of the same name with Lauren Bacall and Edward G. Robinson.

For $65,000, Hendricks bought the boat in 1982. He restored her original seediness and acquired a faded British flag and an appropriately dirty striped undershirt and a sooty cap. He also obtained a special dispensation from the congressional requirement that all "ships" in American coastal trade be American-made. Thus, he can haul nostalgic passengers from the slip behind his 132-room motel at Mile Marker 100.

But he's taken her on other journeys, too. Jim has hauled the five-ton craft on a trailer onto an ocean freighter to participate in a parade of boats across the English Channel to commemorate the World War II evacuation of Dunkirk (all right, so the motor did die in mid-channel and gave Jim's wife, Frankie, some mixed emotions; but that all suits the *Queen*'s image, don't you think?).

However, Jim's love affair doesn't end there. He operates a souvenir shop in his motel built around the boat (Bogie-like undershirts go for forty dollars). He has corresponded many times with Miss Hepburn, who spends much of her time in a house back up Route 1 in Connecticut.

"It's just so much fun," he says. "It's an adventure. It helps my business and Key Largo's, I believe. And I have a business excuse to keep her. I just love boats."

Even without boats, Islamorada (purple island) and its bevy of islands claim to be the sportfishing capital of the world, which is why Carol and Frank Romeka were testing the waters near there the other day. Another retired couple (she was a teacher, he was a sailor and postman), they actually live *and* vacation in Florida, traveling from their home in Jacksonville Beach one block from the ocean all the way down to the Keys and the same ocean. "You know," says Carol, "the ocean is always greener at the other end of the state."

The Romekas make this trek for a week three or four times every year. They are fleeing the phone and the daily chores of retired life, the laundry and shopping, the bills. Usually, they take two days to make the trip down on the state's local back roads, puttering along in the slow lane. "You know," says Carol, "on the Interstate you forget there really are a lot of beautiful places in our country."

The main attraction, however, is the fishing and the carefree hours sitting, baking in the sun, talking idly, speculating on what bait might catch what creature on the next cast. "Down here," says Frank, "you never know what you're going to catch. It's so unpredictable. At home, you know by the weather and the temperature exactly what you're likely to haul in." And if they don't catch anything, so what? No shame, no guilt, no pressure. Maybe tomorrow.

The Romekas fish a different key every day. There's no fishing from the highway bridges, so they pull off on the gravely sand and portage their aluminum lawn chairs to the water's edge down below the elevated roadway. The sun is blinding, reflecting off the sand and the colorful clear water. Their skin is coated in oils, and still drying out. So far this day, they have captured two mangrove snappers—"very tasty," they agree—and tossed a baby barracuda back in. "You never know what those fellows have just eaten," notes Frank, "and, anyway, our freezer is getting full."

In a day or so, they will load up the car with their luggage and frozen fish for future dinners. They'll take U.S. 1 back to the mainland and the Interstate and do seventy all the way home. They are like most vacationers. "Once you start home," says Carol, "your mind gets there ahead of you and you start thinking about all the mail piled up and the bills and restocking the refrigerator and, my God, the lawn. It'll be getting all brown unless we get there soon."

Late in the afternoon, as the tropical sun wanes only slightly, they pack up their rods and chairs and next Tuesday's dinner and head back down 1 toward their rented apartment in Key West. They are delayed only slightly by the closing of U.S. 1 for a high school homecoming parade before that night's big game. The Romekas pass a sugarcane juice stand and a couple of lonely picnic tables, surviving on the old highway. On the horizon, they can see the odd-looking bulbous balloons of Fat Albert and Fat Fanny moored over near a key called Cudjoe (contraction for Cousin Joe). The craft are blimplike military balloons hanging silently over the watery countryside as high-altitude radar platforms with their blips poking the distant sky in search of the incoming drug trade's airborne tentacles.

Down below, some of the victims of the drug trade work off their sentences. Florida is one of the few states that still imposes punishment at hard labor—not the backbreaking, rock-smashing sort of old, with prisoners manacled to each other by metal chains, but tough physical labor repairing roads and collecting litter under contract to the state's Department of Transportation. It's cheaper than hiring civilian labor and maybe makes a point about discipline and hard labor to some medium and minimum prisoners, who likely never held a regular job but now learn that extra hard work gets extra time deducted from their sentence, if their work is judged well.

The Big Pine Key Prison is one of six Florida road prisons. Designed for fifty inmates, it now has sixty-four. "It's better than life in a prison," says a prison official, "but not as good as life in a community release center." With bells announcing each stage of the schedule, the workday begins for the prisoners at 4:30. By 6:45, they are en route to the day's work site. They work until dusk, return to the camp, are strip-searched, fed, and locked up for the night.

Dusk is a bad time for the key deer. They are tiny fellows, none over three feet tall. A big buck weighs perhaps sixty pounds. Dogs chase them. People shoot them. Cars hit them. It's all against the law, of course. Which isn't much help to the key deer, who number barely three hundred today. People think they are just so cute. They draw the little creatures down near the highway by offering traditional deer food such as potato chips and cookies. The animals gather in bundles, begging for the sweets that upset their natural diet and getting caught in garden fences and backyards. And they don't always look before crossing.

Some humane societies estimate that upwards of a million animals die every day on the nation's highways. It's bad enough for those frightened creatures to have to cross the roads that now divide their habitat. But to save space, many of these roadways, including much of Route 1, are divided by so-called Jersey barriers, three-foot cement dividers that create, in effect, a Berlin Wall for any short living thing trying to cross. This turns the pavement into a lethal shooting gallery, with cars as the missiles and real animals as the terrified targets. As a child riding in the backseat, I remember feeling a pang of sadness every time I saw some new crunched carcass along the roadside. I would lie down on the seat and not look out for many miles at a time, so that no more animals would be killed. I used to say a silent little prayer for each one I saw and think that heaven must be getting awfully crowded with these once-wild creatures, because there didn't seem to be much room left for them down here with us.

Cars get about sixty of the key deer every year. Debora Drake sees many of them. She tries to save as many as possible, putting on splints, injecting antibiotics. She loves being a veterinarian, often making sick or damaged little creatures well again and, especially, seeing the loving relief and glow on their owners' faces. Although, a fair number of the animals don't love her being a vet. They remember what happens whenever they smell that disinfectant odor. So they pee on her.

About 40 percent of the Bush Animal Clinic's business concerns cats, another 40 percent dogs (with a few more pit bulls showing up these days). The rest are an impressive assortment of exotic creatures, from the battered deer to iguanas and snakes. The pelicans get fishhooks caught in their beaks, which their mothers never warned them about. They fly into high-tension wires and break a wing or get seriously burned, or worse. They travel down from Jacksonville, like the Romekas, fleeing the frost that burns their feet or, rather, foot; when the birds get cold like that, they stand on one foot and save the other by tucking it up into their insulating feathers. If someone can catch them, Dr. Drake gives them antibiotics.

A lot of people bring in wild animals. One lady recently showed up with an injured mouse, which Dr. Drake tried to save by feeding it diluted dog's milk from a doll's bottle. "We just do it," says the doctor. "It's a life, you know."

There's been a rash of dog poisonings recently. Someone put out antifreeze, which a few animals lapped up. That shut their kidneys down. It was hopeless on some, and they got an overdose of the anesthetic pentathol to avoid further suffering. Other owners desperately approved any treatment, even temporary dog dialysis to get their pet's organs going again. "People here will spend anything," Dr. Drake notes, "to fix up their furry little pals. They are regular members of the family."

The aging dogs and cats, like their owners, come in with blindness, cancer, arthritis, though there's more preventive medicine being sought now, it seems. One woman spent over two thousand dollars at the Bush Clinic and at a specialist, trying to save her red-and-black mutt.

Although there has been some local grumbling about the growing population of resident gays, Dr. Drake is impressed with their love for animals, and, unlike some other community segments, they do pay their bills. The clinic's clientele is an unusual mix of very affluent transients, many of whom travel with their pets, and very poor longtime locals, who don't travel anywhere except up and down Route 1 working or looking for work.

The highway is well cared for in the Keys and seemingly patrolled by one police officer per bridge. There are side streets branching off in each community that wind past pastel-colored homes where rented cars are parked beneath the house, where the first floor normally would be. There are a lot of garden walls with decorative holes carved in the cement, many FOR SALE signs, pert little bridges that climb over some waterway, and platoons of rented mopeds zooming about on the sandy road shoulders.

Outside of Key West, the traffic dwindles drastically as the long dusk wanes. The sunsets out across the Gulf are often firey and colorful. And then the darkness is supreme. Few things are blacker than rural darkness. Punctuated by infrequent all-night yard lights like the ones in the frigid midnights of Maine, the tropical nights are not as long as Maine's, with predawn temperatures sometimes plunging to eighty-six degrees. Many supply trucks make their long runs from the mainland at night, when they actually get up to the fifty-mile-an-hour speed limit because the gawking, inattentive tourists are in bed and the puttering local pickups are back home, too, or perhaps siphoning some liquid sustenance from a neighborhood roadhouse. Viva Zapata's restaurant in Torch Key proudly notes it has a sister facility back up U.S. 1 in Westport, Connecticut.

Key West, sitting out there in the warm waters closer to Havana than Miami, is an unusual bilevel community, though less unusual elsewhere as the economic import of tourism seeps into more areas across the country. On one level, the nation's southernmost city (and, once, Florida's largest) is a normal community with ongoing businesses and long-standing relationships along its palm-studded, one-way streets. On another level, it is a town of transients, where the attractions are touristic, the street directions are baffling, and a long time is a four-night stay. One level wears cameras and tours up and down the usual streets on the sight-seeing trolleys that used to be pulled by mules. The other level doesn't even notice the trolley and passed an ordinance banning sidewalk skateboarding. One level doesn't even bother looking for a parking spot downtown anywhere near the glass-bottom boat dock. The other level drives around and around the blocks in a huge van with out-of-state plates and a fresh FLORIDA KEYS decal on the rear window beneath the sunction-cupped Garfield perpetually climbing the glass. One level says yeah, sure President Truman used to stay here and so did Ernest Hemingway. The other level drives down U.S. 1 and says, hey, look, the Margaret Truman Drop-Off Launderette, and then passes Whitehead Street and says, look, there's Hemingway's house—did you know he wrote *For Whom the Bell Tolls* in those very rooms?

Others note the Geiger house, now a museum available for visiting, where John James Audubon stayed briefly in the 1830s while researching and painting local bird life. Of course, Key West could not waste anything as valuable as Hemingway's birthday (July 21), which has become a five-day annual festival that includes the Hemingway Look-Alike Contest that Tom Cosselman has won, looking more like Ernest Hemingway than Ernest Hemingway (except it is hard to picture the real Hemingway wearing a T-shirt carrying his own portrait).

There are footraces and boat races nearby. There's a lady selling cookies from a cart and food stalls offering chowder, fritters, burgers, and a salad all made out of conch, the shellfish mollusk of which there weren't sup-posed to be any left. Also available for local purchase are the former houses of those large shellfish. Children hold the shells by their ears and listen to the sounds of the ocean's waves coming from within, something that once was part of childhood magic. Adults used to use the shells as horns back in the days when Key West made a pretty good living salvaging the wreckage of passing travelers. Key West's first residents were Bahamians, who brought their pastel gingerbread style of architecture, sometimes shipping an entire house, piece by piece, on a sleek sloop.

Key West's first transients were pirates, who were salvagers in a way; they just didn't wait for the stuff to get lost in the first place. Audubon's Key West host, Captain John Geiger, was a part-time salvager (talk about conflict of interest, his other job was harbor pilot). In Key West, some years like 1846 were so good for ship disasters that an estimated $1.6 million in property floated into local hands, easy as pie. Then they started putting in those damned lighthouses and ruining business. But over on Duval Street, they've salvaged at least a tourist attraction out of the wreckage of the salvage business, turning the Watlington house, Key West's oldest (1829), into the Wreckers' Museum.

For a while last century, the salvage business gave way to sponge fishing (not mentioned now as the most exciting tourist sport), cigar making (smoking draws frowns now, so don't push that), supplying coal to steamships (never mind, out-of-date), providing refuge for Cuban revolutionaries (okay, for now), and, finally, offering a scenic old tourist destination, which was the idea of the helpful Federal Economic Recovery Administration in the 1930s when the hurricane crippled the overseas railroad and helped make Key West bankrupt. Completion of the overseas highway in 1938 made tourism a realistic possibility. And now they stream in by the thousands with their tinted windows closed tight and their ACs on max.

But come to think of it, Key West is still into salvaging. Mel Fisher is an important resident—"a legend in the making," according to the Miami public-relations agency hired to plant as many Key West travel-story ideas as possible in the nation's newspapers (Call 1-800-FLA-KEYS). In two decades of searching, Mel Fisher has recovered some $400 million in gold and silver from the wreckage of a seventeenth-century Spanish galleon that was laden with tons of treasure as well as an unwieldy name, *La Nuestra Señora de Atocha*. Mel Fisher has modestly established the Mel Fisher Maritime Heritage Society's Treasure Museum, which has made available to visitors with tickets the saga of Mel's quest for booty, as well as some selected treasure items for tourist touching.

Key West's original Spanish name was Cayo Hueso, or Bone Island. Even now, the old cemetery gets into the tourism act, attracting visiting photographers with some amazingly cute epitaphs, such as "I Told You I Was Sick" and "At Least I Know Where He's Sleeping Tonight."

Of course, on the other level, there's a standard urban American life afoot, too, in this community. The city of about 25,000 has a substantial artistic and gay community, whose ranks have been particularly ravaged by the modern plague of AIDS, which has hit the big cities all along the highway. In one sense, that health disaster has brought this small-town community together. It has also produced an uncommon public frankness about safe sex and the use of dirty needles. There has been virtually no local controversy, for instance, surrounding one state health officer who makes regular trips on his bicycle, dubbed Randy's Rubber Runs, to distribute condoms from a cookie jar at local stores and guest houses, whose owners leave them on guest pillows at night instead of a mint. Homosexual and heterosexual residents also banded together to raise $700,000 for an AIDS hospice.

Harry Mack remembers when he went for a haircut a half-century ago, *to barber* meant to talk or gossip. And you also got your hair cut. Now, it seems, people don't have the time to spend talking idly. They come in to get their hair cut as infrequently and cheaply as possible, and they are more interested in fads and fashions, which do not interest Harry. He has just figured out this unisex hair business, so he'll cut it that way if you want. And he'll do a few flattops, which are coming back in, although he doesn't like them. He still gets some hippies coming in with their hair down past their shoulders. They pay through the nose; the longer the hair, the higher the price, which starts at six dollars, according to his unposted price list.

Not many mothers seem to bring their little Johnnys out to Harry's on the south side of Route 1. It's just a little shack really. But it's Harry's; he took it over, oh, nearly eight years ago now. He keeps it tidy, though he could do nothing about the palm trees. They once grew all around his barbershop. But some kind of lethal yellow blight seems to be in the air nowadays. It got three of his big old palms. So he cut down two of them, but lopped off the third one at about the six-foot-tall level. Then he painted it like a barber pole. "Yeah," he says. "What do you think of that?"

The pole doesn't twirl. But Harry's is open ten to six Monday through Friday. Mondays are busiest, of course, because he's been closed two days in a row. Harry will not stand for one-day weekends and it's his business. The absolute busiest days are the day after Social Security checks arrive. He's got his longtime local residents coming regularly, the military people, of course, and the retirees who stay in the Keys for weeks at a time; the vacationers don't want to waste their precious time off sitting in a barber chair getting a prickly neck in that heat.

Ever since they widened the bridges, the highway traffic volume seems to be getting worse (the Chamber of Commerce calls it "better"). "They come by here all the time," Harry says, "in their twenty-one-foot-long motor homes, dragging their boats behind. There seem to be more day-trippers, here and gone the same day." A fair number of regular Keys visitors just leave their boats there, stacked up in rental spaces sometimes four-deep down by the marina.

Today's visitors are more upscale, which is nice if they spend some money with you. But for Harry, it means the virtual disappearance of the comfortably dingy neighborhood bar, the type with a few of the old Playmate foldouts stapled on the men's room wall, watching. For so many years, regulars have gathered there of an evening or afternoon to hang out, to gripe, and to relax from another long day in the tropics. Those old places with their aluminum-foil bags of beer nuts are being replaced by slick chrome bars with ferns, no-smoking sections, and little bowls of salty yummies delivered by damsels in short skirts. For this, they charge three bucks for a Bud!

There are more fast-food places, too, the boringly familiar McDonald's and Kentucky Fried Chicken from the mainland, although those Kentucky folks are trying to play down all that fried talk now.

The taxes also have gone up to handle the city services for all the permanent and seasonal residents. Now, Harry says, some $20,000 shack is appraised at $120,000 because across the street some doctor from the Northeast built a quarter-million-dollar mansion. Harry says a lot of his friends have fled the Keys for the quieter corner of Florida up near Georgia or in North Carolina. "I wouldn't mind leaving, too," he says. Though he's not packing.

If he did leave, Harry would have to go back up Route 1. There's not much of the old highway to go on below Harry's. After it passes his shop, U.S. 1 runs close to the water a ways along the beach where much of the newer real estate development has crept because Old Key West is just about full. There's a large sign there, a professionally painted personal expression of appreciation from the Florida Department of Transportation, which contracts its sign making out to the prisoners of the Florida Department of Corrections. The department hasn't got time for commas (the budget crunch, you know). YOU ARE LEAVING KEY WEST FLA., the sign reads. THE BEGINNING OF 1 AND ENDING IN FORT KENT ME. THANK YOU—COME AGAIN. The sign stands by the beach across U.S. 1 from the Ramada Inn, which Fort Kent does not have.

But the highway actually goes on another mile or so into the old city. As Truman Avenue, it passes Big Daddy's Liquors and the Esquire Lounge (TOPLESS!!) near the Hemingway house and garden. Route 1 doesn't quite make it to the little house where Pan American Airways was founded and started international air service from the United States with Flight #1 to Havana on October 28, 1927. That was before most of the highway's current residents were even born, back when ninety miles seemed like a long ways, back before 1 even reached the Keys. That was back before customs agents in Maine looked for cocaine in so many cars, back before Sammy Sanders was postmaster in Calais, before Moody's was serving upside-down roast turkey, before the entire population of Chicago descended on Freeport, Maine, each year. That was before "Pogo," before Jim Grissom's mama kept bees, before New Yorkers killed 2,050 fellow New Yorkers in a year, before Duncan Hines's first recommendation, before Jimmy Bennett dreamed of driving fast, before Okefenokee was a park, before so many prisons, before Interstates, before buildings blocked the beaches, before people shot each other over parking spots, before orange juice cost ten cents. It was also before Pan Am contemplated bankruptcy.

Today, Route 1, the famous transcontinental highway so laden with history and happiness, with trivia and tragedy, scenery and sadness, today the vital roadway simply ends. There is no grand arch, no big sign, no huge bronze plaque. No one standing on *the* spot posing for photos to show the family back home they were actually *There*. Not even a postcard. Just a new book.

Instead, the spot is marked with signs for the Florida Keys Star Printing and Publishing Company, the Girl Friday Temporary Personnel Offices, and the Annex of the Monroe County Courthouse, since the business of taxing, suing, adjudicating, registering, marrying, divorcing, and sentencing Route 1 residents has outgrown the original courthouse.

From the air there, U.S. 1 looks like any other corner in the country, which it is.

ABOUT THE AUTHOR

ANDREW H. MALCOLM IS THE NATIONAL Affairs Correspondent of *The New York Times* and an addicted automobile traveler. For the last quarter century he has roamed the United States and several other countries, filing thousands of stories to that newspaper and writing numerous books on life—and death—in this country.

Currently, he lives in the woods just off U.S. 1 in rural Connecticut with his wife, Connie, two dogs, and too many cats. They spend part of the year in a log house in an isolated corner of Montana just off U.S. 2, driving there, of course.

The author has only one pickup truck (a three-quarter-ton '86 Dodge with 4WD, a 360-cubic inch V-8, a fiberglass cap, and a CB radio). Oh, and he has four children—one son is a reporter in California, one son is a student at the University of Notre Dame, a daughter is in high school, and a third son is a freshman in preschool.

Born in Cleveland, Ohio, Mr. Malcolm grew up in the countryside outside that city before graduating from Culver Military Academy and Northwestern University, where he also earned his master's degree. His books have included *The Canadians, Final Harvest: An American Tragedy, This Far and No More,* and *Someday.*

He reports receiving his driver's license moments after reaching the advanced age of sixteen.

ABOUT THE PHOTOGRAPHER

ROGER STRAUS DIVIDES HIS TIME between photography and book publishing. He and his wife live with three cats on a small island off the Bronx.

INDEX